RICK PETERS

PLUMBING BASICS

Main Street
A division of Sterling Publishing Co., Inc.
New York

Acknowledgements

Butterick Media Production Staff

Photography Editor: Daniel Newberry
Photography: Robert Bonicoro,
    Christopher Vendetta, Brian Kraus
Design: Triad Design Group, Ltd.
Illustrations: Mario Camacho, Ulrike Mönch
Set Building: Tom Perez
Assoc. Art Director: Monica Gaige-Rosensweig

Copy Editor: Barbara Webb
Page Layout: David Joinnides
Indexer: Nan Badgett
Editorial Coordinator: Traci Bosco
Assoc. Managing Editor: Stephanie Marracco
Project Director: Caroline Politi
President: Art Joinnides

Special thanks to Daniel Lenig, Store Manager at Wayne A. Feather plumbing supply house in Emmaus, PA, for poring over the manuscript and gently pointing out technical corrections, and to the production staff at Butterick Media for their continuing support. Also, sincere thanks to my son Will, for his assistance in creating the rough art that was used to make the illustrations for the book. And finally, a heartfelt thanks to my inspiration: Cheryl, Lynne, Will, and Beth. **R.P.**

10   9   8   7   6   5   4   3   2   1

Library of Congress Cataloging-in-Publication Data
Peters, Rick.
 Plumbing Basics/Rick Peters.
  p.cm.
 Includes index.
 ISBN 0-8069-3669-X
 1. Plumbing--Amateurs' manuals. 1. Title.
TH6124.P45 2000
696'.1--dc21   99-086644

ISBN 1-4027-1089-5

Published by Sterling Publishing Company, Inc.
387 Park Avenue South, New York, N.Y. 10016
© 2000 Butterick Company, Inc., Rick Peters
Distributed in Canada by Sterling Publishing
c/o Canadian Manda Group, One Atlantic Avenue, Suite 1(
Toronto, Ontario, Canada M6K 3E7
Distributed in Great Britain by Chrysalis Books
64 Brewery Road, London N7 9NT, England
Distributed in Australia by Capricorn Link (Australia) Pty. L'
P.O. Box 704, Windsor, NSW 2756 Australia

Printed in China

Main Street
A division of Sterling Publishing Co., Inc.
New York

Contents

Introduction

I've always enjoyed helping people. Whether it's showing a novice woodworker how to cut a joint or plane a board, or helping a neighbor repair a leaky faucet, I've found it very satisfying. I think that's why I got into teaching and then eventually into magazine and book publishing: I could reach a larger audience—and help more people. Please don't get me wrong: I don't pretend to know it all. A former student, and good friend, used to say, "Anyone that says they know it all is either a fool or a liar." But I do have a knack for taking a seemingly complex task and breaking it down into bite-sized, digestible nuggets.

That's exactly what I've done for the basics of plumbing in this book. Although it may seem mysterious now, by the time you're done reading this book, it'll seem quite simple. And it is. It's just a matter of wading through the different components and learning how they interact. That's what the first chapter is about—the home plumbing system. It covers everything from the supply, drain, and vent systems to unraveling plumbing codes and permits.

In Chapter 2, I'll take you through the dizzying array of tools and materials that are out there. Again, once you understand the general-purpose and special tools you'll need, and how to use them, your plumbing tasks will be greatly simplified. Likewise, once you've delved into the wide variety of materials, you'll be able to wisely choose CPVC over PVC pipe for that new utility sink you want to install in the basement.

Chapter 3 jumps into working with the various types of pipe and fittings. With the widespread use and acceptance of new materials like PVC and copper pipe, working with pipe has gotten a lot easier. Once you understand how pipe is joined together (especially plastic pipe), you'll feel like a kid with a new Tinkertoy set. It really is that easy.

Murphy's law says that if something is going to break, it'll break at the worst possible time. Chapter 4 covers what to do in an emergency. The two most common—leaks and clogs—are both quite manageable with a little advance knowledge and preparation (like keeping an emergency repair kit on hand).

In Chapter 5, I group faucets into five common types, show you how to figure out which one you've got, and then describe step-by-step how to repair leaks from the spout, around the handle, or from under the base plate.

Chapter 6 is all about the different types of sinks and how to repair or replace them. Integral-countertop, drop-in, and self-rimming kitchen sinks are all covered, along with common accessories like sprayers, garbage disposers, and dishwashers.

Everyday toilet adjustments and repairs are described in Chapter 7. Everything from stopping a toilet from running with a simple adjustment (you adjust the mysterious device called a ballcock) to stopping floor leaks by replacing a wax ring. Although not the most glamorous job, everyone in your home will appreciate your ability to fix a malfunctioning toilet.

Finally, Chapter 8 details how to replace or install several common plumbing fixtures: showerheads, spouts, sinks, and faucets, to name a few. Again, all it takes is a little knowledge, the right parts, and a few hours of your time.

Most plumbing repairs and remodeling chores are not that difficult if you take them one step at time. Every task in this book is broken down into simple discrete steps and is illustrated with detailed drawings and clear photographs. *Plumbing Basics* is a step-by-step approach—a working handbook and a quick reference guide for the most common repairs you'll be faced with. I hope it helps you in your do-it-yourself endeavors.

Rick Peters
Spring 2000

Chapter 1

Home Plumbing Systems

Before taking on any plumbing job, it's important to have a basic understanding of how the plumbing system in your home works. In this chapter, I'll start by describing the three main systems in your home that work together: the supply system, which provides fresh water to all the fixtures in the home; the waste system, which removes liquid and solid waste; and the vent system, which allows the waste system to drain properly. Critical to both the drain and vent systems are the traps—I'll explain how they work and walk you through the different types.

Then I'll help you unravel the mysterious and often misunderstood plumbing codes, rules, and permits that are essential to safe plumbing. And finally, I'll show you how to check your home to make sure the existing plumbing is sized and installed correctly. No fancy calculations are required—just count fixtures, take a few measurements, and check the charts in this chapter to see whether the existing plumbing is up to snuff. (Keep in mind that this is a quick check—call in a plumbing professional or a building inspector for a comprehensive inspection.)

If you have an older home that has had numerous families, chances are, over the years, some of the plumbing was altered by a previous owner. And unless these alterations were inspected and approved by the local building department, you could have an unsafe system or one that doesn't work properly. (Even a new home can have a poorly installed plumbing system.)

If the water pressure or drains in your home seem sluggish, the problem may be improperly sized or installed pipes in the system and not a clog or malfunctioning fixture. I always suggest to homeowners that they take the time to check their system before trouble occurs and correct any discrepancies. That way when you do encounter an emergency, you'll have confidence that your system is sound and not the cause of the problem. (*See Chapter 4* for emergency repairs.)

Supply Systems

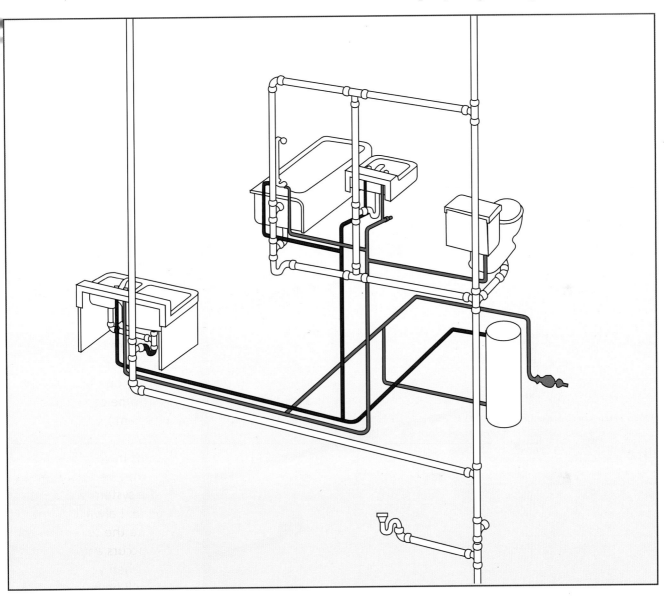

The supply system in your home directs pressurized water to the plumbing fixtures. Fresh water is supplied by the local water utility or from a private well; the pressure comes from the city's pumping stations or from the well pump, respectively. Regardless of the source, the water flows through a main shutoff valve (and through a water meter if supplied by a utility) and then to the hot water heater.

From there, both hot and cold water branch out to various parts of the home. These lines are referred to as branch lines. Lines that transport water up to a second floor or up into a fixture are called risers. All lines terminate in some sort of valve that when opened will allow water to flow. Valves such as tub, sink, and shower faucets operate manually. Other valves, like those in toilets and refrigerator icemakers, are automatic.

Waste Systems

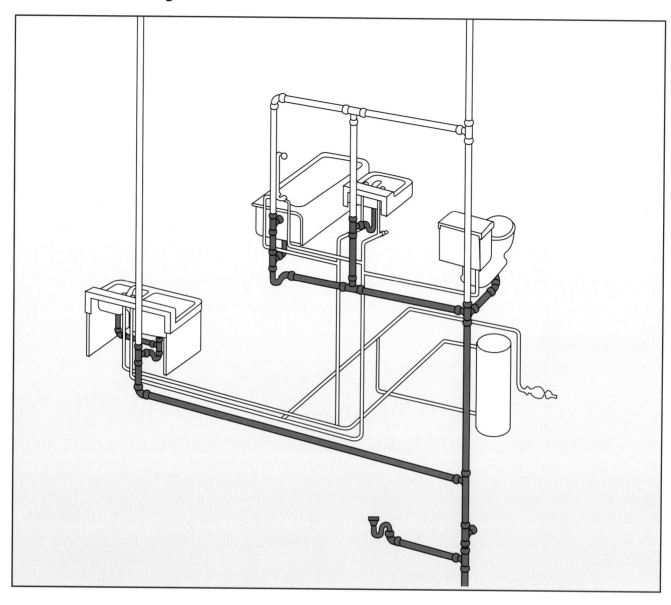

The waste system transports solid and liquid waste out of the home. The system relies on gravity to move the waste water from sinks, showers, tubs, and toilets out of the fixture and into a waste line (often called the soil stack) that empties into the municipal sewer or a private septic tank. In between every fixture and the waste line is a trap—basically a curved section of pipe that captures or "traps" water (toilets have built-in traps). The trap fills with sufficient water to form an airtight seal to prevent sewer gas from entering the home. Since the trap is curved, solid waste can build up in it and clog the line. Fortunately, they're easy to remove and clean out (see pages 50–51). You'll also find other points in the system where clean-outs have been added so that you can deal effectively with the inevitable clogs.

Vent Systems

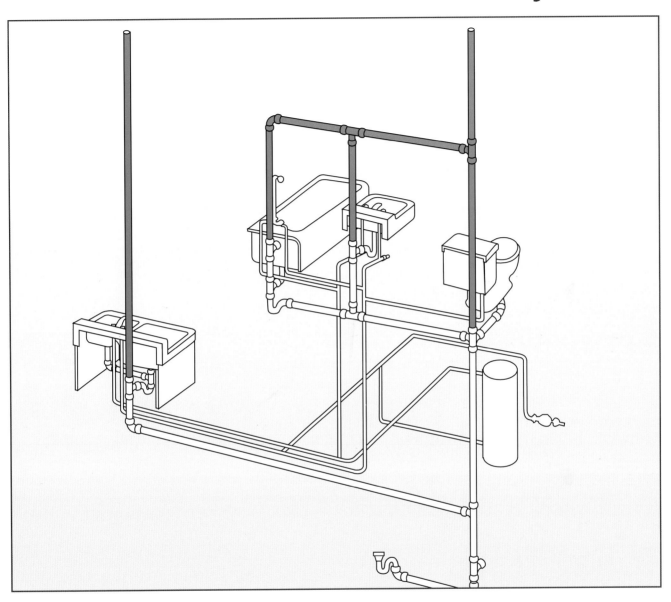

The vent system is often the most misunderstood part of the home plumbing system. Many novice DIYers think that all a fixture needs to work is water and a drain. Not so; the third element—the vent—is also critical. In a nutshell, the vent does two important jobs. First, it allows the wastewater in the drain system to flow freely. Second, it prevents siphoning, which can pull the water out of traps, allowing sewer gas into the home. In both cases, a vent does its job by allowing fresh air to flow into the drain system the same way the second hole (or vent) in a gas line can allow the gas to flow out freely. Vents are connected along each fixture's drain line past the trap. Depending on the number of fixtures and layout of your home, you may have a single vent, or multiple vents. In addition to allowing fresh air in, vents also allow sewer gas to flow out of the home and harmlessly up through the roof vent.

Traps

P-Traps are the most common traps used in plumbing today. A P-trap is designed for drain lines coming out of a wall (instead of up through the floor, as in S-traps; *see below*). Shaped like a P lying face-down, P-traps are usually installed under sinks, bathtubs, and showers. They consist of two main parts: a trap arm, and a removable U-bend. Slip nuts join the parts together and allow for easy removal and cleaning.

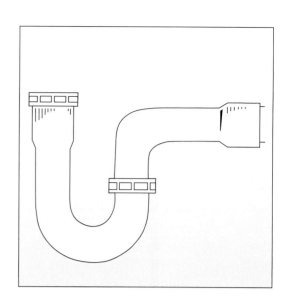

S-Traps were common when drain lines came up through the floor. Because of their design, S-traps are prone to self-siphoning, which can cause the seal to fail, allowing sewer gas into the home. It should be no surprise then to learn that S-traps are prohibited by most codes in new construction (you can still replace an S-trap with an S-trap in your home, however). Since S-traps can be found in many older homes, you can still buy replacement S-traps at most plumbing supply and hardware stores.

Toilet Traps

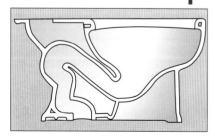

Siphon Jet Trap The trap at the rear has a hole in the bottom that sends a jet of water into it, creating a siphoning action when flushed.

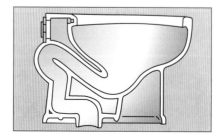

Reverse-Trap Similar to a washdown, but the trap is at the rear of the bowl. This makes the bowl longer and provides for a quieter flush.

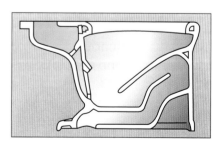

Washdown Trap Inexpensive and simple, but noisy. The trap is at the front of the bowl and is flushed by streams of water draining from the rim.

Plumbing Codes

Although they may seem to be a nuisance, plumbing codes are written and enforced to protect you and your family. Plumbing codes protect everyone's health, safety, and welfare. Virtually every city, county, and state has adopted a plumbing code based on one of three national plumbing codes: the BOCA National Plumbing Code, the Standard Plumbing Code (SPC), and the Uniform Plumbing Code (UPC). In Canada, the basis is the National Plumbing Code.

Each of the national codes covers a wide variety of subjects (*see the box below right*)—everything from the design and installation of plumbing systems in new construction to alterations to existing homes. The individual sections of the code are very specific in describing allowable materials, proper pipe and fixture sizing, and installation layouts and measurements.

For example, say you need to replace the trap for your shower. The only size available at the local hardware store is 1½" in diameter. Can you use this? No, code specifies that the traps for showers be at least 2" in diameter. Why? Because of the volume of water they need to handle. A smaller trap wouldn't handle the flow and could back up and overflow. Instances like this make it obvious why it's so important to follow the applicable code. The various codes are all based on the learning experiences of thousands of plumbers—professionals who've learned from their mistakes and shared their knowledge so future plumbing installations will be safe and worry-free.

Part of the confusion associated with plumbing codes is that what is perfectly acceptable in one part of the country may be prohibited in another part. This is particularly significant regarding allowed materials. It's important to note that your local plumbing code supersedes the national code that it amends. The only way to make sure is to check the local code—often a photocopied list of amendments to one of the national codes. It's often available free or for a small fee from your local building department; the national code your community uses can usually be found as a reference item at the local public library.

Plumbing codes are laws—not suggestions or a set of directions for you to follow—that are enforced by local inspectors. Failure to comply with these can result in fines and other penalties. Homeowners who install their own plumbing must follow the code and are subject to the same fines as a licensed plumber. Although this may sound harsh—it's no different from enforcing traffic laws—they're there to protect everyone.

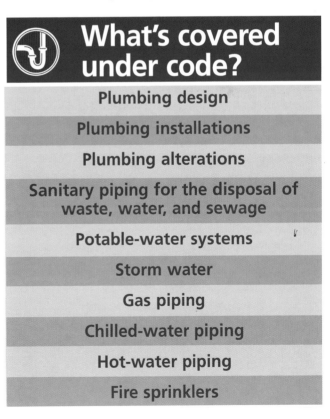

What's covered under code?

Plumbing design

Plumbing installations

Plumbing alterations

Sanitary piping for the disposal of waste, water, and sewage

Potable-water systems

Storm water

Gas piping

Chilled-water piping

Hot-water piping

Fire sprinklers

Permits

All plumbing codes require you to apply for a permit before doing substantial work on your plumbing system. Not all work requires a permit; *see the sidebar below.* Applying for a permit informs the building department that you're planning to alter the plumbing in your home. It's their job to make sure the work is done to code. Can you do the work without a permit? Legally, no. Illegally, yes—but you run the risk of an improper installation that can both affect the health of your family and cause potential problems when you sell your home. Don't do it—take the time to make sure the job is done right.

Obtaining a permit will require you to thoughtfully plan out the job—don't expect the building department to do it for you. Once the plan is approved and a permit is issued, you can begin work. At specific stages of the job, a plumbing inspector will come out to see how things are progressing. (Don't hesitate to ask detailed questions about the inspection procedure—you'll be glad you did.)

If you're adding new lines (roughing-in work), the inspector will want to look at and test the integrity of the new system. Because of this, the new lines must be fully accessible—you can't

cover up or conceal any of the work. If you do, you'll have to remove the obstruction. When the job is complete, the inspector may come out again to check the finish work—that is, the installation of the plumbing fixtures.

Unit ratings for fixtures

Fixture	Unit Rating
Bathtub	2
Clothes Washer	2
Dishwasher	2
Kitchen Sink	2
Shower	2
Toilet	3
Utility Sink	2
Vanity Sink	1

DO I NEED A PERMIT?

Simple repairs like fixing a leak, clearing clogged traps, pipes or sewer lines, and repairing faucets or valves do not require a permit. (Most of the projects in this book won't require a permit.) You will, however, need a permit anytime you consider adding to or changing existing plumbing, running new plumbing, or upgrading substandard plumbing. If in doubt, call your local building department for a ruling. In some cases, they may need to come out and take a look before deciding whether you need a permit or not.

Checking Your System

If you're having problems with your current plumbing system (such as poor water pressure or slow drains) or you're planning on upgrading your plumbing system, it's a good idea to check to see whether your water distribution system is sufficient. To do this, start by determining the total demand on your system.

Total demand is calculated in unit fixtures, by adding up the unit ratings for all the fixtures in your house; *see the chart on the opposite page.* Then measure the total length of the water distribution pipes in your house. With these two numbers, *check the chart below* to determine the correct size of your distribution pipe (note that this will also depend on the size of the service pipe coming in from the street).

If the distribution pipe is undersized and needs to be upgraded, consult a plumbing professional. Even if the distribution pipes are sized correctly, you may still have water flow problems if the supply lines aren't sized properly. The supply lines connect the fixture to the distribution pipes. Have a plumbing professional check to make sure that your supply lines are sized right. If they're not the right size, they're simple to replace; for more on this, *see page 45.*

Sizing of water-distribution pipe

Size of Service Pipe from Street	Size of Distribution Pipe from Meter	Maximum Length Needed for Total Fixture Units					
		40	60	80	100	150	200
¾"	½"	9	8	7	6	5	4
¾"	¾"	27	23	19	17	14	11
¾"	1"	44	40	36	33	28	23
1"	1"	60	47	41	36	30	25
1"	1¼"	102	87	76	67	52	44

Common Requirements

In addition to distribution pipes and supply tubing, there are a number of other common requirements you can check. The first thing to find out is whether the valves and traps in your existing plumbing system are causing any problems. Check to make sure that your system has the correct valves in place. You should have a full-bore gate valve or ball valve on both the street side and the house side of your water meter. (For more on valves, *see page 27*.)

You should also find one of these valves at the inlet of your water heater and, if you have a water-based heating system, on the inlet to the boiler. In addition to this, all of the individual plumbing fixtures in your house should have their own shutoff valves. If they don't, I'd suggest installing them—you'll be glad you did when an emergency arises.

Traps that are improperly sized can also lead to problems with the waste system. The type of fixture will determine the size of the trap. Check the chart *below* to make sure you've got the correct trap for the fixture; if not, replace it with the proper size. It's also important to check and make sure that the distance from the trap to the vent doesn't exceed the maximum length. If it does, you'll likely experience drain problems due to poor venting.

Trap size and distance for fixtures

Fixture	Fixture Units	Minimum Trap Size	Maximum Trap to Vent
Bar Sink	1	1½"	3½'
Bathtub	2	1½"	3½'
Clothes Washer	2	2"	5'
Kitchen Sink	2	1½"	3½'
Shower	2	2"	5'
Toilet	3	3"	6'
Utility Sink	2	1½"	3½'
Vanity Sink	1	1¼"	2½'

If all your traps are the correct size but the drain system is still sluggish, you can check to make sure that the horizontal and vertical drain pipes are sized correctly. The sizing of each different pipe depends on its load, determined by the total fixture units that it serves. Add the number of fixture units for each pipe (*see page 12*), and then *check the chart at right* to make sure the pipe is sized correctly.

It's also critical that horizontal pipes are sloped the appropriate amount. Pipes less than 3" in diameter should slope ¼" per foot toward the main drain. Pipes larger than this should slope ⅛" per foot.

Another factor in a properly functioning drain system is appropriately sized vent piping. *Check the chart at right* to make sure that the vent for the drainpipe it serves is the correct size and the correct distance away from the drain.

Finally, it's a good idea to make sure that all the plumbing lines in your home are supported properly. If they aren't, they may sag and strain the connecting joints, causing leaks. Water surging through pipes that aren't supported properly can also cause the pipes to rattle. *See the chart at right* to determine the correct support intervals for the type of pipe installed in your home.

Vertical & horizontal drainpipe sizing

Pipe Size	Max. Units for Vertical Drain	Max. Units for Horizontal Drain
1¼"	1	1
1½"	2	1
2"	16	8
2½"	32	14
3"	48	35
4"	256	216

Critical distances for vent pipes

Size of Fixture Drain	Minimum Vent Size	Maximum Trap-to-Vent
1¼"	1¼"	2½"
1½"	1¼"	3½"
2"	1½"	5"
3"	2"	6"
4"	3"	10"

Support requirements for pipe

Pipe Material	Maximum Horizontal Spacing
Cast iron	5' (may be 10' where 10' lengths of pipe are installed)
Copper tubing and pipe	6' for 1¼"-diameter & smaller 10' for 1½"-diameter & larger
Plastic DWV	4'
Plastic pipe, rigid	3'
Plastic pipe, flexible	32"
Threaded steel	10' for ¾"-diameter & smaller 12' for 1"-diameter & larger

Chapter 2
Plumbing Tools & Materials

Without a doubt, one of the most daunting tasks involved with any plumbing repair or remodeling job is wading through the myriad tools and materials that line the aisles and shelves in hardware stores and home centers. I can't tell you how many folks I've seen over the years wandering aimlessly down the aisles with their eyes glazed over from an overabundance of choice—kind of like a kid in a candy store. The difference here is a wrong choice in tools or materials could mean a repair job gone awry, often resulting in an emergency call to the local plumber. But this doesn't have to happen. All it takes is a basic understanding of what tools and materials are out there, what they're used for, and when.

In this chapter, I'll start by describing the basic kit of general-purpose tools you'll need for almost any plumbing job—the stuff every homeowner should have in their toolbox (*pages 17–18*). Then I'll discuss the specialty tools that will make a seemingly impossible task possible (*pages 19–21*). (We've all come across a situation where we figured out that there must be some

special tool to get the job done—we just didn't know what it was, or how to use it.) In plumbing, this can be anything from a spud wrench to a flexible tubing bender. Then on to the different types of sealants that we'll depend on to create watertight seals (*page 21*).

Finally, I'll take you through the vast array of plumbing materials for you to choose from. Everything from the advantages, disadvantages, and uses of the different types of pipe (*pages 22–23*) and the gaggle of fittings available—copper (*page 24*), plastic (*page 25*), and drain-waste-vent (*page 26*)—to transition fittings that allow you to connect dissimilar materials. Armed with this information, you'll be able to confidently make tool and material choices for almost any plumbing task.

General-Purpose Tools

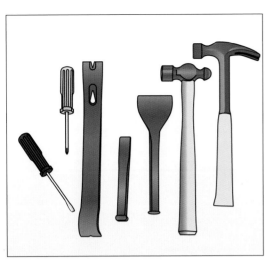

Demolition Many of the plumbing jobs you'll tackle will require some demolition work; that is, tearing out an old section of wall, flooring, or cabinet. You'll find the following tools useful for this type of work. *From left to right:* screwdrivers for general dismantling; a pry bar for pulling out stubborn boards and fixtures; a cold chisel or set of inexpensive chisels for chopping out holes in walls or flooring; and ball-peen or claw hammers.

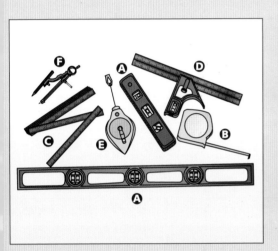

Measuring and Layout One of the most critical steps in any plumbing remodeling job is measuring and laying out the placement of piping and fixtures. The tools shown should be in every homeowner's toolbox: a 3' (A) and torpedo level (A); a 25' tape measure (B); a folding rule for short, accurate measurements (C); a combination square to check for right angles and for general layout (D); a chalk line for striking long layout lines on walls and floors (E); and a compass to draw circles (F).

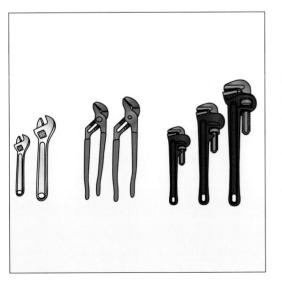

Gripping Tools At the heart of a plumber's toolbox are different gripping tools for loosening and tightening nuts, caps, fittings, valves stems, etc. Although it's helpful to have different sizes, you'll need at least one each of the following tools. *From left to right:* an adjustable wrench for smaller nuts and fittings; a set of channel-type pliers for larger nuts and fittings; and a pipe wrench for working on galvanized or cast-iron pipe.

Power Tools Used primarily for plumbing remodeling, power tools can make quick work of many tedious jobs. *Clockwise from top left:* a reciprocating saw for demolition work and cutting pipe; an electric drill with a ½" chuck for large-diameter holes; a cordless drill with a ⅜" chuck for smaller holes; and a saber saw for cutting access holes, holes in countertops, and any curved cut that requires precision.

Unclogging Tools When an emergency clog occurs, these are the tools to reach for. *From left to right:* a plunger with a flanged lip for clearing clogs in toilets and kitchen sinks; a standard plunger to clear most lavatory sinks; a closet auger primarily used to retrieve objects stuck in a toilet's trap; a crank snake for augering through stubborn clogs; and a long hand snake for clearing long runs or vent stacks.

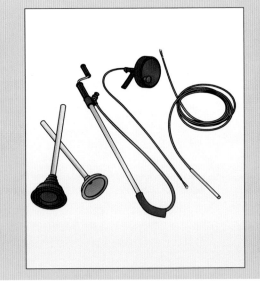

Protective Gear As with any home improvement work, it's important to protect yourself by wearing appropriate protective gear. Keep the following on hand. *Clockwise from top left:* knee pads not only to cushion your knees, but also to protect them; ear muffs or plugs for when working with power tools; safety goggles for work involving power tools, demolition, or joining pipes; and rubber or leather gloves to protect your hands when working with sharp edges or hot pipes.

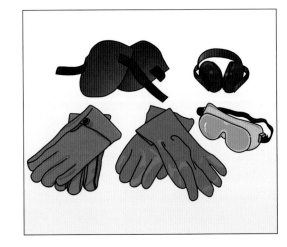

Specialty Tools

Many of the simple plumbing repairs described in this book can be accomplished with the general-purpose tools listed on *pages 17 and 18.* But for more complicated repairs, and much of the remodeling work, you'll need a couple of different specialty tools. These are the tools for cutting pipe and the wrenches designed for a specific task (such as a seat wrench used to remove and install faucet seats). Most can be purchased at any hardware store or building center for a few dollars. Alternatively, you can borrow them from a neighbor who is handy— or if there's a friendly plumbing supply store in your area, they may lend you the tool, or you can rent it for a nominal charge.

Cutting Many plumbing tasks will require you to cut and fit pipe—either plastic or copper. The cutting tools to have handy are (*from top to bottom and left to right*): a hacksaw for general-purpose cutting of just about anything; a pipe saw for cutting larger plastic pipe; a mini-hacksaw for cutting pipe in tight spaces; a standard tubing cutter to cut copper pipe cleanly; a pair of plastic tubing pliers for clean cuts on small-diameter plastic pipe; and a compact tubing cutter for cutting copper in tight spaces.

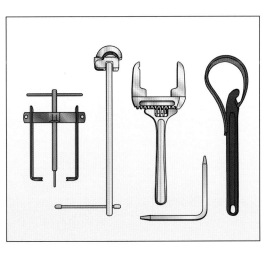

Wrenches I'm sure you've experienced a situation where there was only one tool that could get the job done. In plumbing, these tools are (*from left to right*): a faucet handle puller for removing stubborn handles; a basin wrench for reaching up and loosening faucet nuts; a spud wrench to remove the spud washer on a toilet; an L-shaped seat wrench to remove or install seats on faucets; and a strap wrench for working on large-diameter pipe.

Tools for Working with Copper

Regardless of whether you're working with flexible or rigid copper pipe, you'll need a few special tools for joining or "sweating" the pipe, and a couple of different tools for shaping and bending pipe. In addition to the tools shown here, you'll also need some of the protective gear shown on *page 18*. Since joining copper pipe (as described on *pages 36–39*) uses molten solder to join or "sweat" the pipes together, it's imperative that you protect yourself. Make sure to wear leather gloves and eye protection whenever working around open flames and heated parts. Rigid copper, although easily joined, isn't designed to bend—instead you use fittings to wrap around obstacles (*see page 24*). Flexible copper, on the other hand, can be shaped and bent—but only with the tools listed *below.*

Sweating Joining together sections of copper pipe requires special tools for cleaning and heating the joint: a source of heat— emery cloth to clean the ends of the pipe (A); a wire brush for cleaning the insides of the fittings (B); a propane torch (C); a spool of lead-free solder (D); flux (E), which both cleans the joint and helps the solder to flow; a flux brush to apply it (F); a spark igniter (G); and a heat-resistant cloth to protect the area surrounding where you're sweating the pipe (H).

Flexible-Copper Tools Although there are only two specialty tools for working with flexible copper, they're a must. First, for flare fittings *(see page 43),* you'll need a flaring tool to flare the pipe end out like a small trumpet. You can bend flexible tubing by hand, but don't risk it—it kinks very easily. Instead use a tubing bender; it slips over the tubing and prevents kinks as you shape the curve. Also shown *(far right)* is a conduit bender, which can be used to put slight bends in larger-diameter flexible copper.

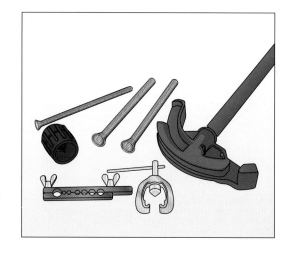

Sealants

I'm sure whoever coined the phrase "An ounce of prevention is worth a pound of cure" wasn't thinking about plumbing sealants—but it certainly applies here. Besides using the correct fixtures and components, there is no better way to ensure a watertight joint than to use a sealant—and use it properly. *See the chart below* to match up the sealant to the type of job you're working on. Although some plumbing manufacturers make fixtures that don't require a sealant, I prefer to add a turn or two of Telfon tape to male threads, or a dab of pipe-joint compound to female threads, as added insurance. I'd rather take the time up front and do the job right than have to come back later when a leak occurs.

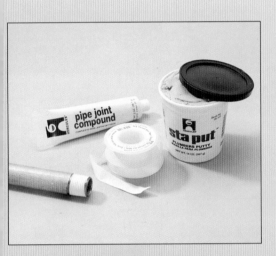

Sealant Types The mainstays of leak prevention, *(clockwise from right to left):* plumber's putty, Teflon tape, pipe-joint compound, and silicone. Plumber's putty is a soft, doughlike substance available in cans or plastic tubs. Teflon tape is a thin membrane that wraps easily around pipe threads. Pipe-joint compound, similar to toothpaste, also comes in squeeze tubes for ease of application. Waterproof and flexible silicone—an excellent sealant for around toilets and vanities—is also available in squeeze tubes but is most often found in tubes for caulking guns.

Common plumbing sealants

Sealing Material	Uses
Teflon tape	Seals male pipe threads against pressure leaks. Comes in ½" and ¾" widths.
Pipe-joint compound	Seals female pipe threads against pressure leaks. Available in cans with a screw-on cap/brush, or in squeeze tubes.
Plumber's putty	Prevents gravity leaks around faucets, sinks, strainers, and overflows. Comes in inexpensive plastic containers.
Silicone caulk	Creates a watertight seal around fixtures such as sinks, toilets, showers, and tubs. Available in cartridges or squeeze tubes.

Supply & Waste Pipe

Common Supply and Waste Piping (*from left to right*): Cast iron, typically used in older homes for the main waste system, PVC (polyvinyl chloride), the modern equivalent of cast iron, ABS (acrylonitrile butadiene styrene), approved for cold water use only, CPVC (chlorinated polyvinyl chloride), a common replacement for galvanized iron or copper pipe supply lines, and galvanized iron, the standard in most older homes for hot and cold supply. *See the chart below* for the specifics on each type of pipe.

Supply and waste pipe

Material	Uses	Form
ABS	Drainpipes and traps	10', 20' lengths
Cast iron	Main drain waste	5', 10' lengths
CPVC	Hot and cold water supply	10', 20' lengths
Galvanized	Drainpipes; hot and cold water supply	1" to 1' nipples, custom lengths up to 20'
PVC	Drainpipes; vent pipes	10', 20' lengths
Brass	Valves and exposed drain traps and pipes	various short lengths
Chromed Copper	Supply tubing for fixtures	12", 20", 30" lengths
Flexible Copper	Hot and cold water supply	30', 60', 100' coils
PB	Hot and cold water supply	25', 100' coils
Rigid Copper	Hot and cold water supply	10', 20' lengths

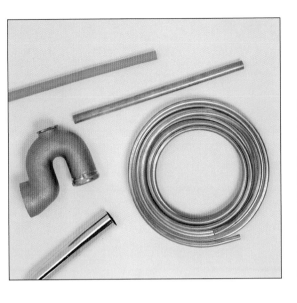

Common Supply and Waste Piping—continued (*clockwise from top left*): PB (polybutylene), plastic pipe for hot and cold supply lines that's not approved in all areas, rigid copper, the standard for hot and cold supply lines in new construction, flexible copper for supply line installations where joints need to be kept to a minimum, chromed copper used mainly where supply lines are in plain sight, and brass pipe for exposed drain and traps. *See the chart below* for the specifics on each type of pipe.

Advantages	Disadvantages
Lightweight, low cost, easy to cut and assemble	Cold-water approved only, restricted in some areas
Very strong, quiet	Difficult to work with
Hot-water approved, low cost, easy to cut and assemble	More expensive than PVC
Strong, quiet	Expensive due to threaded joints, susceptible to corrosion/scaling
Lightweight, low cost, easy to cut and assemble	Cold-water approved only, damages more easily than other piping
Heavy, durable, and attractive	Expensive
Attractive surface where appearance is important	Scratches easily, expensive
Requires fewer fittings, can use compression and flare fittings, bends easily	More expensive than rigid copper
Installs with fewer fittings, no freeze damage, not susceptible to water hammer	Difficult to join, not allowed by all codes
Fast and easy to assemble, resists corrosion, durable	Susceptible to water hammer and freeze damage

Copper Fittings

Copper fittings are most commonly used in a supply system to direct hot and cold water to fixtures. Larger fittings and pipe can be used for small-diameter drainpipe. There are dozens of fittings available that will allow you to connect pipe into almost any configuration. Fittings are sized for a specific-diameter pipe. For example, you'll find ½" and ¾" fittings in almost every hardware store. Some of the more commonly used fittings are shown in the drawing at *right. From left to right and top to bottom,* they are: a coupling with internal stop, a 45° elbow, a 90° elbow, a 90° street elbow, a 45° street elbow, a T fitting, an end cap, and a male adapter.

"Street": fittings have one male end and one female end. This allows you to connect fittings together without having to join them with short sections of pipe (or nipples).

Ninety-degree elbows allow piping to turn a sharp corner; 45° elbows work well for more gradual turns. A T fitting is used for lines that intersect, or if you need to tap off a line and direct water to another fixture. A coupling with stop allows you to join pieces of pipe together and is used a lot in repair work for splicing together repaired section of pipe. This type of fitting has an internal stop that prevents you from pushing the pipe too far into the fitting.

Male adapters and female adapters are used when you need to connect a threaded fitting on the end of the pipe, such as a shutoff valve for a toilet or sink faucet. Tube or end caps are used to seal off a pipeline—whether for testing, for emergency repair work, or for a remodeling job. For step-by-step instructions on working with copper pipe and fittings, *see page 36.*

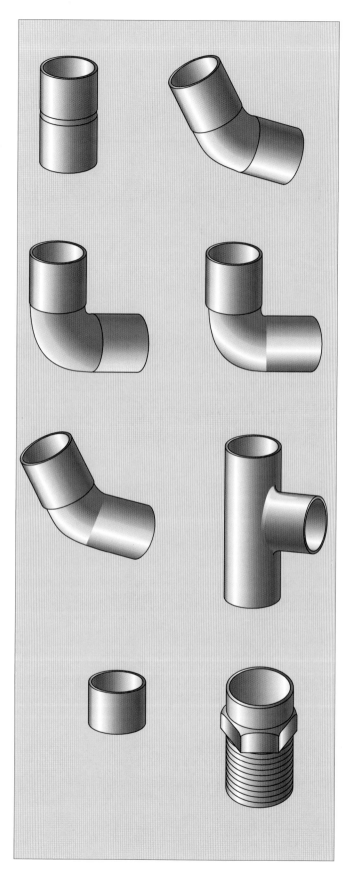

CPVC Fittings

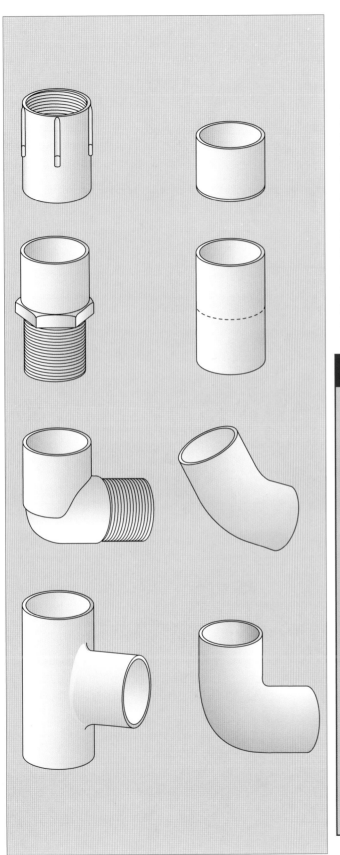

Just like copper fittings, CPVC fittings can be used for hot and cold supply systems. And just like copper, you can find dozens of fittings that will allow you to navigate pipe around almost anything. The fittings shown in the drawing at *left* are some of the most commonly used. *From left to right and top to bottom,* they are: a female adapter, an end cap, a male adapter, a coupling, a 90° male adapter, a 45° L, a T fitting, and a 90° street L. The use of each of these fittings is described on *page 24.* For a step-by-step description of working with PVC pipe and fittings, *see page 31.*

BUYING TIPS

Unlike copper fittings, which can be reheated, separated, and reused, plastic fittings are joined with cement. Once set up, the joint cannot be taken apart. Even when you take your time and are careful assembling a section of pipe, it's possible for something to go wrong—typically a pipe or fitting ends up out of alignment. That's why I always purchase at least two of every fitting I'll need—and for fittings that I'll be using a lot (like 90° elbows), I'll pick up a handful.

There's nothing more frustrating than having to stop midway through a plumbing job (especially when you've shut off water to the entire house) to run down to the local hardware store—or worse, drive across town to the building center—to pick up a $.49 part (that is, if they even have it in stock). When the job is done, you can keep the extras on hand for the next job (or in case of an emergency) or simply return them.

DWV Fittings

Because they're designed to handle a greater volume of water and waste than other fittings, drain, waste, and vent fittings are larger and beefier in construction. DWV fittings can be purchased in either white PVC (polyvinyl chloride) or black ABS (acrylonitrile butadiene styrene), with PVC being the more common. (Note: You should always match the fitting material to the type of pipe you're using; that is, you can't mix and match PVC with ABS—they use different cements.) DWV fittings have gentle curves instead of sharp corners and are designed to keep wastewater flowing downhill. Because of this, they are commonly called sanitary fittings.

Some of the most often used fittings are shown in the drawing at *right. From left to right and top to bottom,* they are: a 22½° L, a 22½° fitting L (a fitting L is similar to a street L—one end is smaller, to fit inside another fitting), a 90° fitting L, a 90° L, a 45° L, a T fitting, and a 45° Y. Many of the L-type fittings are available in a "long-turn" bend that provides for an even gentler and more gradual curve.

The variety of DWV fittings is almost staggering: You should be able to find a fitting for almost any situation. Sanitary branches are designed for places where two or more waste lines converge. There are also wide selections of clean-outs that provide access to the waste stack for augering or cleaning out clogs *(see page 56).* For step-by-step instructions on working with plastic pipe, *see page 31.*

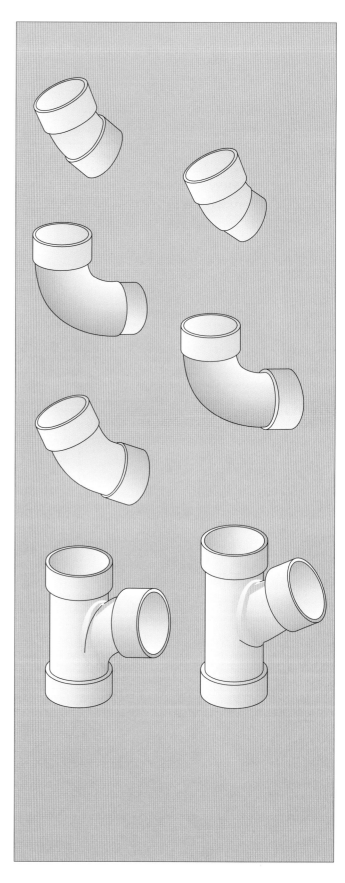

Supply Valves

Supply valves are available in either metal or plastic. Regardless of how they work internally, they all serve the same purpose: to open or close a water line. You'll find one of these on the water line coming into your house—the main shutoff valve. In older homes, this is most likely a gate valve. Gate valves use a wedge-shaped brass "gate" to block off the flow of water. This type of gate is not as reliable as the globe valve shown *below.* It cannot be repaired if it doesn't fully stop water flow, and it should be replaced if the opportunity arises.

Every fixture in your home should have its own shutoff valve. If this isn't the case, I recommend spending part of a weekend installing one for every fixture. When an emergency occurs, or when it's simply time to make a repair, you (and your family) will appreciate the fact that you can do the job without turning off the water for all or part of the house.

Common Valves (*Clockwise from top right*): a brass drain valve used to drain water out a section of pipe; a plastic fixture shutoff valve used to independently turn off water to a fixture; a brass gate valve commonly found as the main shutoff for water coming into the house; and a brass hose bib typically affixed to the exterior of a house for a hose, or in a laundry room for connection to a clothes washer.

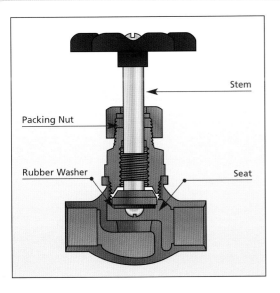

Stem

Packing Nut

Rubber Washer

Seat

Cross Section A globe valve works similarly to a compression faucet (*see page 70*). When the handle is turned clockwise to the off position, the stem moves down into the body until a rubber washer on the end of the stem meshes with a "seat" to block the flow of water. This type of valve is very reliable and is easy to repair. If it doesn't fully stop the flow of water, the stem can be removed and the washer replaced.

Transition Fittings

Copper-to-Galvanized You can connect copper pipe to galvanized pipe, but only if you use a dielectric union. If you don't use a dielectric union, the two dissimilar metals will react, causing an electrochemical reaction to occur, resulting in corrosion. A plastic spacer inside the union keeps the metals from direct contact. One end of the dielectric union threads onto the galvanized pipe; the other is soldered onto the copper pipe.

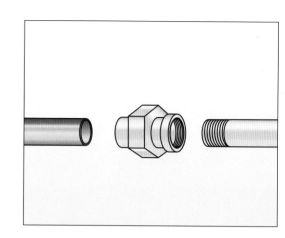

Plastic-to-Galvanized You can connect plastic pipe to galvanized pipe by using male and female plastic threaded adapters. Plastic adapters are joined to the plastic pipe with the appropriate cement and then threaded onto the galvanized pipe. The threads of the pipe should be wrapped with Teflon tape to guarantee a good seal. (For more on Teflon tape, *see page 21*.)

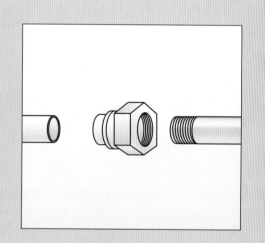

Copper-to-Plastic Copper pipe can be joined to plastic pipe with a plastic double-ended compression fitting. A nut and a rubber or plastic compression ring is slipped over the end of each pipe. When the nuts are threaded onto the fitting and tightened, the ring compresses, forming a watertight seal. For more on compression fittings, *see page 42*.

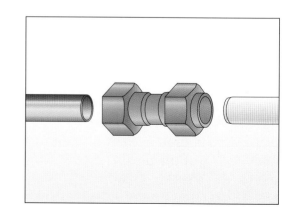

Measuring Pipe & Fittings

Inside Diameter In order to buy the correct pipe or fittings, you'll first need to identify the size of pipe you'll be connecting to. Pipe is sized based on the inside diameter (I.D.). If you have access to a cut section of the pipe, hold a tape measure or ruler across the widest part of the pipe. The inside diameter, or nominal size, is the distance from one inner wall to the other.

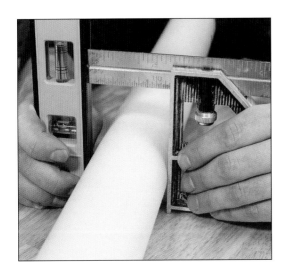

Outside Diameter In situations where you don't have access to a cut pipe, you can measure the outside diameter and use the chart at right to determine the nominal I.D. of the pipe. The simplest way I've found to measure the O.D. accurately is to wrap a try square around the pipe, as shown, and butt it up against a torpedo level. Then read directly off the square and check the chart for the I.D.

Typical pipe dimensions

Material	Nominal I.D.	Approx. O.D.	Approx. Socket Depth
CAST IRON	2"	2¼"	2½"
	3"	3¼"	2¾"
	4"	4¼"	3"
COPPER	¼"	⅜"	5/16"
	⅜"	½"	⅜"
	½"	⅝"	½"
	¾"	⅞"	¾"
	1"	1⅛"	15/16"
	1¼"	1⅜"	1"
	1½"	1⅝"	1⅛"
PLASTIC	½"	⅞"	½"
	¾"	1⅛"	⅝"
	1"	1⅜"	¾"
	1¼"	1⅝"	11/16"
	1½"	1⅞"	11/16"
	2"	2⅜"	¾"
	3"	3⅜"	1½"
	4"	4⅜"	1¾"
THREADED	⅜"	⅝"	⅜"
	½"	¾"	½"
	¾"	1"	9/16"
	1"	1¼"	11/16"
	1¼"	1½"	11/16"
	1½"	1¾"	11/16"
	2"	2¼"	¾"

Chapter 3
Working with Pipe

In the not too distant past, most homeowners shied away from working with pipe—and for good reason: The old galvanized and cast-iron systems that were common in just about everyone's homes required the services of a plumber. When called upon, they often brandished exotic materials like molten lead and oakum (for working with cast iron) and strange-looking machines such as mules and tripods (thread-cutting machines for galvanized pipe). Not the kind of stuff you'll likely have down in the basement.

But all this has changed with the advent of new pipe materials. Although cast-iron and galvanized pipe are still used today (though less and less), they're rapidly being replaced with easier-to-work materials such as plastic (PVC and CPVC) and copper pipe.

In this chapter, I'll start by describing step-by-step how to work with three common types of pipe, starting with the easiest: plastic, either PVC or CPVC (see page 31), then copper pipe (page 36), and finally galvanized iron (page 46). Then

I'll cover installing and removing brass valves (page 40), compression fittings (page 42), flare fittings (page 43), and how to work with flexible tubing—both copper and reinforced plastic (page 44).

The type of pipe you work with will largely depend on what's installed in your home. In most cases, you should replace existing pipe with similar pipe. If you're ever in doubt, see your local building inspector to check code in your area. In some instances, the local code will be very specific as to what type of pipe can be replaced with what. If you're adding new plumbing, however, your options are wider. Here again, check your local code for a listing of the choices for material options.

Joining Plastic Pipe

By far the simplest pipe to work with, plastic pipe can be cut and joined with ease. There are a number of ways to cut it: with a tubing cutter or pliers, with a power miter saw, or with a hacksaw (don't forget to wear safety glasses). The important thing is to make sure the cut is square. If it's not, the joint will be weak and the seal questionable at best. Also, since all plastic pipe is not the same (*see page 22*), make sure you're joining like materials (PVC to PVC, etc.). Dissimilar materials are usually not compatible and will degrade over time.

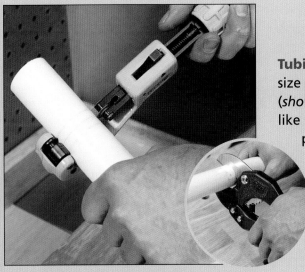

Tubing Cutters or Pliers Depending on the type and size of plastic pipe you're cutting, a standard tubing cutter (*shown*) will offer varying degrees of success. Harder pipe, like CPVC, cuts well and leaves a clean crisp edge. Softer pipe, such as PE, compresses under the cutting wheel of a tubing cutter and is best sheared off with a pair of tubing pliers (*inset*). If you're cutting plastic pipe that's over 2" in diameter, you'll be better off using a power miter saw (*below*) or a hacksaw and a miter box.

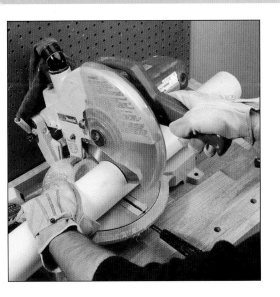

Power Miter Saw Equipped with a carbide-tooth saw blade, a power miter saw will slice through even the thickest CPVC like butter. I replaced the entire waste/vent stack in a three-story 1890 Victorian house in one day with the help of this saw. It's particularly handy when you need to trim off just a bit—⅛" or so—something that's extremely difficult to do with a hacksaw or tubing cutter.

Hacksaw and Miter Box If you don't have a lot of pipe to run (or if you can't get your hands on a power miter saw), you can get by just fine with a hacksaw and a miter box. The miter box doesn't need to be fancy—something shop-made is fine as long as it'll guarantee a straight cut. Use your hand, a wedge, or a clamp to hold the pipe tight against the side of the miter box when cutting.

1 Deburr with a knife Regardless of what method you use to cut plastic pipe, you'll need to clean up the cut ends. Run a utility knife around the inside edge to remove any burrs, which could clog up an aerator, or even worse, restrict water flow. To make it easier to slip pipe and fittings together, some plumbers also cut a small chamfer or bevel around the outside edge of the pipe with their knife.

2 Test the fit and mark I can't overemphasize the importance of this step. Testing the fit of your cut pipe and fittings can save you the trouble and expense of redoing a job. Once cement is applied, the joint sets up almost immediately. After you've assembled your pipe in place, draw an alignment mark across each joint. This way you can take the pieces apart, apply cement, and reassemble them with complete confidence.

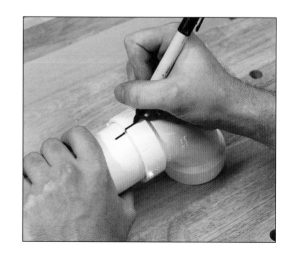

3 **Apply primer to pipe** Depending on the pipe, you may need to prime it before applying cement. ABS doesn't require this; PVC does. Primer does two things for PVC pipe. First, it removes dirt and grease. Second, it actually softens the pipe surface slightly so the cement can form a better bond. Most primers have a built-in dauber that eliminates the need for a brush. As you swab on the primer, make sure to cover the entire joint area.

4 **Apply cement to pipe** All plastic pipe cement isn't the same—you'll need to buy cement for the specific material you're joining—there are even different cements for PVC versus CPVC. Wipe a generous amount around the pipe with the built-in dauber or a brush. But don't get carried away—solvents in the cement "melt" the material to form the bond; excess cement in a thin-wall pipe can create a future disaster.

5 **Apply cement to fitting** After you've applied cement to the pipe, wipe on a thin coat of cement inside the fitting. Here again, don't get carried away. Although it may be tempting to apply primer and cement to a number of pieces at one time—don't. The cement sets up so fast that I suggest working on one joint at a time. It's slower, but in the long run this will save you the trouble of having to redo work.

6 **Roughly position pipe** Working quickly, insert the pipe in the fitting until it bottoms out. Adjust the pipe so the marks you made in Step 2 are roughly one quarter turn apart. You should see some excess cement oozing out near the joint. This is okay; it's a good indicator that you've used enough cement. If it's dripping, you've used a bit much. Don't worry; you'll clean this up after the next step.

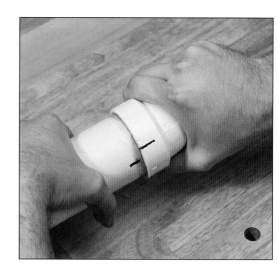

7 **Twist to spread cement** With the pipe and fitting assembled, twist the pipe one quarter turn until the marks are aligned. Twisting the pipe like this helps spread the cement inside the joint to ensure a good bond. In some types of pipe, there is a tendency for the pipe/fitting joint to move after assembly. To prevent this, simply hold the joint together with modest hand pressure for a minute before moving on to the next joint.

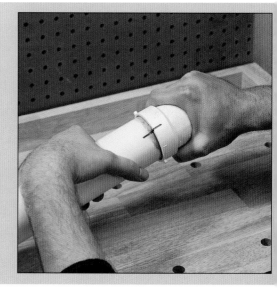

8 **Wipe off excess** After you've joined the pipe and fitting, wipe off any excess cement with a clean, soft rag. This isn't just for neatness; remember, the solvents in the cement will soften the pipe—anywhere it makes contact. Removing excess cement is added insurance that you're less likely to encounter problems with a damaged pipe down the road.

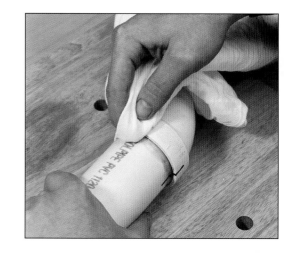

Disassembling Plastic Pipe

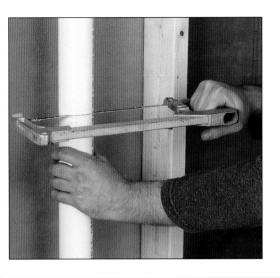

1 **Cut apart with a saw** Whether you're doing a repair, refitting a joint, or adding a new line, you'll often need to cut into an existing run of plastic pipe. With the water off and a bucket and towel handy, cut through the pipe with a hacksaw, mini-hacksaw, or reciprocating saw. Since you'll likely be rejoining to this pipe, take the time to make the cut as square as possible.

2 **Add union to reconnect** After you've cut through the pipe, remove any damaged or unwanted sections. Clean and deburr the ends of the pipe to which you'll be joining fittings. Using the procedure described on *pages 32–34,* add a union or tee fitting to the end of the pipe.

PREVENTING ELECTRICAL PROBLEMS

The electrical system in your home may be grounded through your metal water pipes. If you are planning on replacing a section of metal pipe with easier-to-work-with plastic, you'll be breaking the ground loop. This can cause a wide variety of electrical and safety problems. To prevent this from happening, pick up a couple of grounding clamps and a length of stout copper wire at your local hardware store. Attach each clamp securely to the metal pipe. Connecting a wire between the clamps will complete the circuit and eliminate any potential troubles.

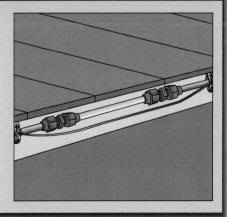

Joining Copper Pipe

Copper pipe is only slightly more complicated to work than plastic pipe. The big difference is how the pipe and fittings are joined together. Unlike plastic pipe, which uses a solvent-based cement, copper pipe and fittings are joined with solder. The joint is "sweated" or heated with a propane torch until it's hot enough to melt solder. Solder is then flowed into the joint to create a strong, watertight seal. To do this type of work, you'll need the specialty tools used for sweating pipe shown on *page 20.*

1 **Position cutter and twist** The most common tool for cutting copper pipe is the tubing cutter. To use a tubing cutter, open the jaws until it can be slipped over the pipe. Adjust its position until the cutting wheel is directly over a mark made on the pipe that indicates the desired length. Tighten the knob and rotate the tubing cutter around the pipe to begin cutting.

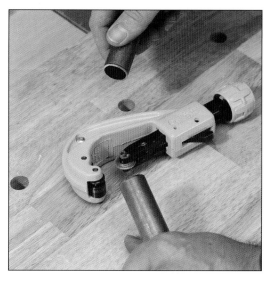

2 **Cut through the pipe** Continue alternately tightening the knob and rotating the tubing cutter until you've cut through the pipe. Depending on the wall thickness of the pipe and your hand strength, this may take anywhere from 5 to 20 turns. You may find it helpful to wear gloves for a better grip, as the pipe tends to twist as the cutter is rotated.

3 **Ream out inside of pipe** The cutting action of the tubing cutter creates a lip (basically a continuous burr) all the way around the inside of the pipe. This lip will restrict water flow and needs to be removed with a reamer. A couple of twists is all it takes. You can use a hand reamer, *as shown,* or the retractable reamer that comes built-in on most tubing cutters.

4 **Deburr outside of pipe** After you've removed any burrs from the inside of the pipe, you can turn your attention to the outside. A strip of emery cloth is just the ticket here. Wrap it around the pipe and give the pipe a few twists. This does two things. First, it removes any burrs that would cause problems fitting the joint together. Second, it removes oxidation that would prevent the solder from forming a solid bond.

5 **Clean inside fittings** Just as you did with the outside of the pipe, you need to remove any oxidation from the inside of the fittings. A special round wire brush makes this an easy task. In a pinch, you can use a brass brush or else wrap some emery cloth around your finger, insert it in the fitting, and twist.

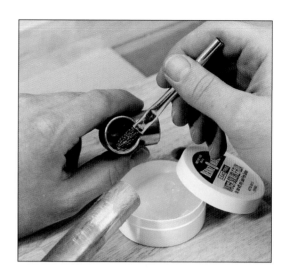

6 **Apply flux** There's one more thing to do before you assemble the joint—apply flux. When heat is applied, flux burns off any remaining oxidation and also helps the solder flow into the joint. Apply a generous glob to both the pipe and the fitting with a flux brush. Two quick tips: Trim the bristles of the brush to about ¼" long for better control, and store the used brush in a toothbrush holder to keep it clean.

7 **Assemble joint** Push the flux-covered parts together until they bottom out. If you've made alignment marks, twist the pipe until they're aligned. When working on existing plumbing, it may help to insert a small ball of bread into the pipe (*inset*). The bread stops any remaining water in the pipe from preventing a good solder joint, and it'll dissolve when the water is turned back on.

8 **Adjust torch flame** The heat for "sweating" pipes is provided by an inexpensive propane torch. Light the torch and adjust the flame. What you're looking for here is a steady, blue flame at the heart, with a bit of yellow surrounding it. Spend the extra couple of bucks for a spark igniter. You won't have to go looking for matches, and since they keep your hands farther away from the flame, they're safer to use.

9 **Heat the joint** Apply the flame about an inch away from the base of the fitting. The more metal you're working with, the longer it will take. Move the flame from the fitting onto the pipe in gentle, sweeping motions, concentrating on the fitting. The flux will begin to sputter and spurt as it melts and cleans the joint. If you're working on existing plumbing, protect nearby surfaces with heat-resistant cloth or a piece of sheet metal.

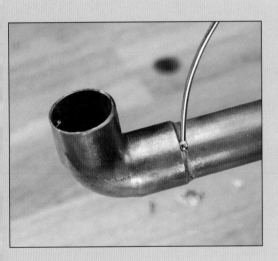

10 **Apply solder** Remove the heat, and touch a length of solder to the joint. If it melts, it's ready to go. If it sticks to the fitting, apply more heat and try again later. Continue applying solder to the joint—it will wick into the fitting until it's full. You should see a continuous bead of molten solder around the entire joint. Don't go wild here: "More is better" does not apply in this case.

11 **Wipe away excess** Wearing leather gloves and using a clean, soft rag, wipe away any excess solder before it has a chance to set up (do this quickly as the molten solder will set up in just a few seconds). Excess solder can trap flux, dirt, and debris in pockets, which can eventually weaken the joint. It's best to remove it.

Joining Brass Valves

Brass valves can be joined to copper pipe using virtually the same procedure as described for joining copper pipe (*pages 36–39*). There are two things that are different, however. First, the large metal mass of the brass fixture will require a longer heat-up time before solder can be flowed successfully. Second, inside most valves is a plastic washer that creates the seal to block off the flow of water. This seal will melt from the prolonged heat application if you don't take measures to prevent this from happening (*see below*).

1 **Protect the seal** Although you might think the best way to protect a washer in a valve is to remove the valve stem, it's not. The stem is sealed by the manufacturer. If you break the seal, odds are the valve will leak. A simpler solution is to prevent heat from building up in the first place by allowing it to escape. Just back out the valve stem, loosen the valve nut, and pull it away from the body. The heat will travel up the valve stem the way smoke goes up a chimney.

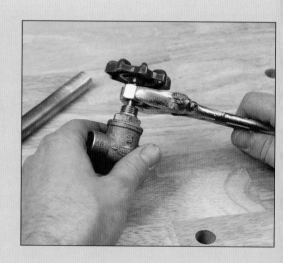

2 **Apply heat** Once the valve is prepared as described in Step 1, it can be joined to the pipe. Have patience with your propane torch: There's a lot of metal to heat up. After you've sweated all the joints, let the valve cool to the touch before threading the valve nut back in place and tightening it.

Disassembling Copper and Brass

Unlike the cemented joints made with plastic pipe, the soldered joints of copper pipe can be disassembled by reheating them and pulling them apart. If this is impractical because of where you're working (under a cabinet, in a crawl space, etc.), you can instead cut the pipe and rejoin sections with a union. Just like plastic, you can cut the pipe with a hacksaw, but I recommend a tubing cutter for a clean, square cut. If there isn't sufficient clearance to spin a standard cutter, there's a "mini" version that fits in the palm of your hand for those extra-tight situations.

1 Apply heat To disassemble copper pipe and fittings, apply heat to the fitting until the solder in the joint turns molten. If you can't readily see this, remove the heat and touch a length of unspooled solder to the fitting. If it melts without the end of the solder sticking to the pipe, it's hot enough. Don't forget to protect surrounding areas with heat-resistant cloth or sheet metal before applying heat.

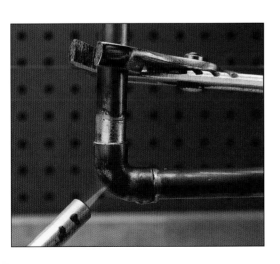

2 Pull apart Wearing leather gloves and using slip-joint pliers, pull the pipe and fitting apart. If you're planning on reusing pipe or fittings, you'll need to remove excess solder. When hardened, excess solder makes it difficult to assemble new fittings and pipe. Hold the fitting or pipe with a pair of channel-type pliers and apply heat until the solder is molten. Quickly, before it sets, wipe off the excess with a clean, soft rag. Once it's cooled, emery cloth can smooth out any remaining solder.

Joining Compression Fittings

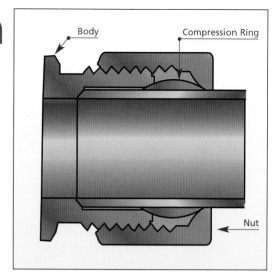

Body   Compression Ring

Nut

Compression fittings are used where you'll want to be able to take things apart easily; for example, supply lines to fixtures that need maintenance, such as faucets, dishwashers, and icemakers. A compression fitting consists of three parts: a body, a compression ring, and a compression nut. The nut and ferrule slip over the pipe, which is inserted into the fitting. Tightening the nut compresses the ferrule into the fitting, creating a watertight joint.

1 Position the ring and seal After you've cut your copper pipe to length with a tubing cutter or hacksaw, ream out the ends and then slip the compression nut and ferrule over the pipe. Slide them both down the pipe and then apply a bit of pipe-joint compound around the ferrule. Although compression fittings should work without a sealant, consider using it as added insurance against leaks.

2 Tighten to compress the ring Insert the pipe into the fitting until it bottoms out. Then slide the compression nut in place and thread it onto the fitting. Wrap a couple turns of Teflon tape around the threads of the fitting, and then use an adjustable wrench to secure the nut and compress the ferrule. Here again, more is not better. The threads on compression fittings are very fine and strip easily if overtightened—take it easy.

Joining Flare Fittings

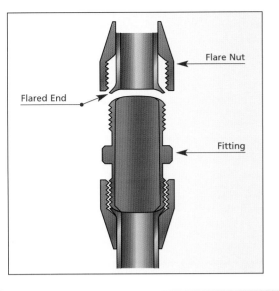

Flare Nut

Flared End

Fitting

Although flare fittings are most commonly used for flexible copper gas lines, they can be used to connect flexible water supply lines as long as the connections won't be concealed. The trick to getting a good seal with a flare fitting has less to do with the fitting, and more to do with the flare you create on the end of the pipe with a special flaring tool. The flared end of the pipe mates with a beveled edge inside the fitting. When a flare nut that slips over the pipe is tightened, a watertight seal is formed.

1 **Flare the ends of the pipe** Start by slipping a flare nut over the pipe. Then select the appropriate hole in the flaring tool and clamp it on the end of the pipe as shown. Fit the yoke of the flaring tool around the base, and center it in the pipe. Gradually tighten the yoke to form the flare. Go slowly here and don't overtighten, or you may crack the flare and have to start all over again.

2 **Assemble** Back the yoke out of the pipe and release the pipe from the base of the flaring tool. Position the flared end of the pipe over the fitting, and slide the flare nut into place. Thread it on by hand and then tighten the joint with a pair of adjustable wrenches: one on the fitting, the other on the flare nut.

Joining Flexible Tubing

There are a number of flexible tubing products available that can make your adventures in plumbing a lot more enjoyable. Flexible copper tubing, or "rolled" tubing, is great for long runs where sweating connections would be inconvenient. It's inherent flexibility also makes it a good choice for connecting supply lines to icemakers. As a refrigerator is moved in and out for cleaning or servicing, a flexible copper supply line offers some give and take. Flexible supply and waste lines, available in a variety of lengths and diameters, make reconnecting fixtures a snap; *see the sidebar on the opposite page.*

1 Uncoil the tubing Flexible tubing comes in coils, often in boxes. The first thing to do is uncoil the desired length. The best way I've found to do this is to straighten it out as it comes out of the box. Pull out a couple inches' worth at a time, straighten it, and continue. The factory ends of the tubing are often out of round, so I suggest trimming the first few inches off before cutting the tubing to final length.

2 Bend with a tubing bender If you need to bend the tubing into a curve, especially a tight one, don't do it without using a coil-spring tubing bender (*shown here*). Slip the bender on the pipe and exert gentle hand pressure until the desired curve is achieved. If you try to bend this type of tubing without a bender, you'll most likely create a kink in the pipe, which will eventually fail and cause a leak.

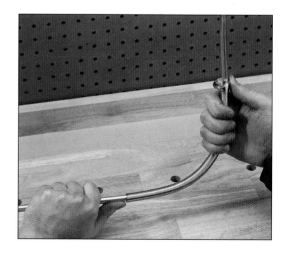

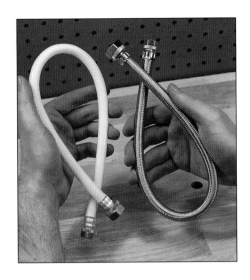

Flexible Supply Lines Although chromed copper supply lines are somewhat flexible, braided metal and vinyl mesh supply lines like the ones shown make fixture hookup a breeze. They're available in a variety of preset lengths, complete with captive connecting nuts. When purchasing flexible supply lines, you're better off long than short; any excess can bend to one side. If they're way too long, they can even be looped.

QUICK CONNECTS USING FLEXIBLE TUBING

Flexible supply and waste lines are a DIYer's dream. No specialty tools are required—just an adjustable wrench or a pair of channel-type pliers, and you're in business. Flexible supply lines (*right*) typically run between shutoff valves and fixtures—usually in tight, cramped spaces. Any flexibility in these situations is appreciated.

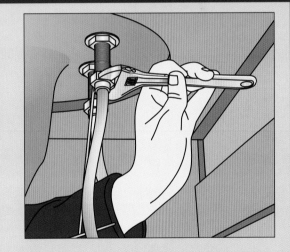

Flexible waste lines (*right*) are often located next to their smaller cousins, but they connect sinks and tubs to the waste stack. Most flexible waste lines use plastic compression fittings to make the connection. Just slide the compression nut onto the fitting and hand-tighten—it's that simple. Flexible waste lines are particularly useful when replacing sinks where the drain hole is in a different location.

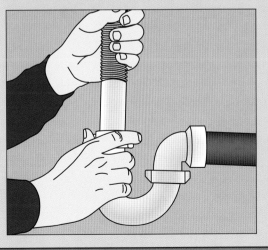

Joining Galvanized Pipe

1 **Loosen with pipe wrench** Since it's threaded on both ends, galvanized pipe can't be removed at random. You have to start at the end of a run or at a union (whichever is closer), and work back to the section you want to replace. To dismantle the pipe, position a wrench on the pipe and another on the fitting so their jaws are facing opposite directions, as shown. Move the wrenches toward the jaw opening to loosen.

2 **Or cut and remove** In cases where the union or end of run is a considerable distance away, you may be better off just cutting the old pipe out with a hacksaw or a reciprocating saw fitted with a metal-cutting blade. Once this pipe is removed, you'll have to install a union here so that you can reconnect the system.

TIPS FOR LOOSENING STUBBORN PIPE

Here are two tips to help loosen a tenacious pipe. The first thing to do is apply penetrating oil to the joint. Let it sit, and have at it again. If it still won't budge, try heating the joint with a hair dryer, heat gun, or propane torch. The heat will help the metal to expand or contract and break the joint.

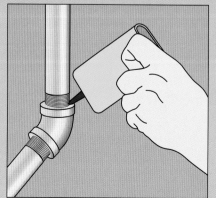
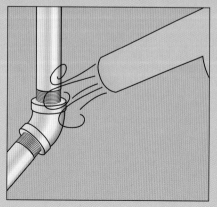

3 **Clean the threads** Once you've got the old pipe removed, you'll need to measure for a replacement pipe. Take this dimension to your local hardware store or plumbing supply house to have a replacement pipe cut to length and threaded. Before you reassemble the system, take the time to clean off all the threads with a wire brush. Clean inside the fitting as well—anyplace bits of metal and dirt can be captured.

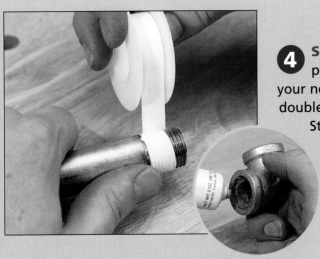

4 **Seal the threads** The last thing you want to happen after you've gone to this much trouble is for your newly installed pipe joints to spring a leak. I use a double insurance system that has rarely let me down. Start by wrapping the male threads with a couple turns of Teflon tape. Then apply a generous dollop of pipe-joint compound to the female threads of the fittings (*inset*).

5 **Reassemble** Now you can reassemble the system in the reverse order that you took it apart. Starting at the replacement pipe, work back to either the union or the end of run. Take your time and work with one joint at a time. Thread the parts together by hand, and finish tightening them with the pipe wrenches. Be careful not to overtighten or you'll crack the fittings.

Chapter 4
Emergency Repairs

Most plumbing emergencies can be classified into one of two categories: clogs and leaks. In this chapter I'll start out by describing how to handle common clogs: bathroom sinks (*page 49*), kitchen sinks (*page 54*), shower and tub drains (*page 58*) and finally, toilets (*page 59*). Then on to repairing minor and major leaks (*pages 60 and 62, respectively*).

Many clogs can be cleared with a plunger. I recommend keeping both types on hand: a non-flanged plunger for bathroom sinks and some tubs, and a flanged plunger for toilets and kitchen sinks. I'd also recommend gathering together a small plumbing emergency kit consisting of a closet auger (for stubborn toilet clogs), a hand auger (for clearing drain lines), and a pipe-repair kit (available at most hardware stores) for pinhole leaks and burst pipes. You'll be glad you've prepared in advance when trouble strikes (and Murphy's Law says it will at the absolute most inconvenient time—like a dinner party with the boss or a family gathering around the holidays).

Clogs, while unpleasant, don't create quite the excitement as water spewing out of a wall, a floor, or a ceiling. In these instances, it's imperative that everyone in your home knows not only where the main water shutoff is, but also how to stop the flow of water. Make sure there's always a clear path to the shutoff valve so that it's easy to get to.

If you do encounter a minor or major leak, the first task is to find its source. In some cases, it's apparent: The water is pouring out of a pipe that burst. At other times you'll have to do a little quick detective work to trace it down. Start by looking for obvious clues like a picture newly installed on a wall with a large nail (yikes) or an overflowing sink or toilet. If it's not obvious, try tracing the water flow back to the nearest fixture. For leaks behind an enclosed wall, you'll have to cut a hole to access the pipe. Keep the hole as small as possible, and save the section you cut out—you may be able to reuse it when you patch the wall.

Unclogging a Bathroom Sink

Clogs in a bathroom sink are usually caused by a buildup of soap, hair, toothpaste residue, etc. Although the knee-jerk reaction to a clog is to use a chemical drain cleaner—don't. Most commercial brands aren't strong enough to clear clogs. Professional-strength versions do a better job, but there's risk involved: The chemical reaction that clears the clog can also generate sufficient heat to damage your pipes, particularly plastic traps and clean-outs. Instead, reach for a plunger. If the problem persists, you can remove the trap to clear debris, or if that doesn't work, use a snake on the trap arm. If the problem is in the waste stack, I'd suggest calling in a professional drain-cleaning service.

1 Remove the drain stopper Before you can effectively plunge a bathroom sink, you'll need to remove the drain stopper. There are two versions of drain stoppers: one type simply lifts directly out; with the other type, you'll have to remove a pivot rod that connects to the pop-up mechanism (*see page 84*). Clean the drain stopper and set it aside.

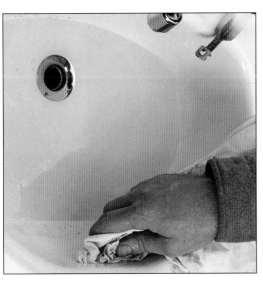

2 Block the overflow The other thing to do to prepare a sink for plunging is to block off the overflow hole. If you don't, your plunging efforts will be in vain. That's because the water you plunge will simply bounce off the clog and travel back up the overflow. To prevent this, block the overflow hole or slot by stuffing a damp rag or a sponge in it. For rigorous plunging, you may need a helper to hold the rag in place.

3 **Fill with water and plunge** A plunger requires a good seal to force water down the drain. The seal depends on how the cup fits against the basin and the water level. Keep the water in the sink above the cup and try folding back the cup lip to get a better seal. Plunge with a series of quick, strong strokes, giving the plunger an extra-hard shove on the last stroke. Repeat five or six times to clear the clog.

4 **Remove the clean-out** If plunging didn't work, the next step is to clean out the trap. Some traps have a clean-out plug on the bottom of the bend to provide access to the trap without having to remove it. With a bucket under the trap, unscrew the plug and poke a screwdriver or a wooden spoon handle into the clean-out. Stir around to break up the debris. Try plunging again to clear the clog.

5 **Remove the trap** If your trap doesn't have a clean-out, you'll need to remove the trap. Start by placing a bucket under the trap, and have a couple of shop towels handy for spills. Use a pair of channel-type pliers to loosen the slip nuts on both ends of the trap. To keep from scratching the chrome nuts, wrap a couple layers of masking tape or duct tape around the serrated jaws of the pliers. Once loose, unscrew the nuts by hand and lift out the trap.

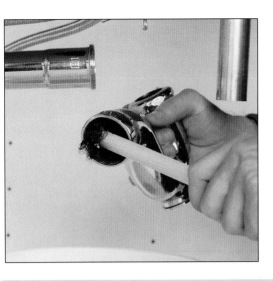

6 **Clear any trap debris** Shake the trap over the bucket to dislodge any loose debris. If the clog is compacted down in the trap, force it out with a screwdriver or the handle of a wooden spoon. Work in both directions as needed to clear the debris. If there's a lot of sludge built up in the bottom of the trap, force a piece of cloth through the trap to clean it out.

7 **Insert the auger** If you pull off the trap and discover that there's no debris, the clog in most likely in the drain line leading to the sanitary T, which connects to the waste stack. A hand auger can often be used to successfully clear this type of clog. Start by inserting the end of the auger cable into the drain opening until you feel resistance.

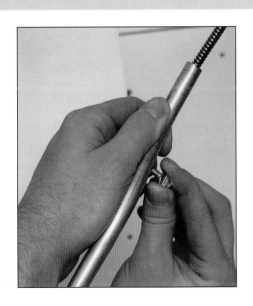

8 **Lock the auger** The resistance you feel is one of two things: Either it's the clog, or it's a bend in the drainpipe. When you hit the resistance, leave around 5" or 6" of cable exposed and turn the thumbscrew where the cable exits the auger to lock the cable in place.

9 **Crank the arm** Now twist the arm on the hand auger in a clockwise direction. This causes the cable to spin. If the resistance is a bend in the drain line, the spinning cable will be able to navigate past it. If the resistance is solid, you've found the clog.

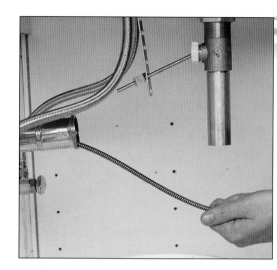

10 **Push the auger** Continue twisting the arm while exerting pressure on the cable. If you feel continuous resistance but can still slowly advance the cable, the clog is probably the result of a buildup of soap. Continue cranking and exerting pressure until the end of the cable bores through the clog.

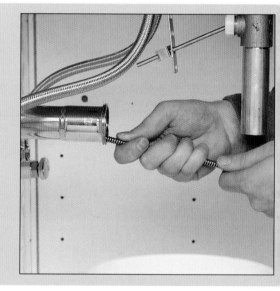

11 **Pull the auger** In some cases, the clog is caused by a buildup of hair, or something accidentally flushed down the drain, like a bit of sponge or a piece of jewelry. If this is the case, you may be able to snag it with the end of the cable and pull the clog back out of the drain line. Once you've hooked onto the object, release the auger lock and crank the handle as you pull out the obstruction.

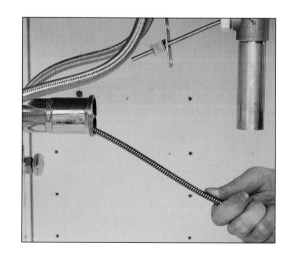

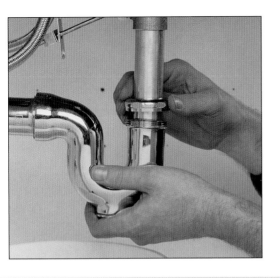

12 Reassemble Once you've cleared the clog, you can reassemble the drain line. Hold the trap in position and hand-tighten the slip nuts. Use the channel-type pliers to snug the nuts tight. If you've removed the clean-out plug, wrap a couple layers of Teflon tape around it and thread it back into the bend. Reinstall the drain stopper, connecting it to the pop-up linkage if necessary.

13 Flush with water Turn the water on gradually to flush out any remaining debris. Let the water run for a while to make sure you've cleared the clog. If after all this the water still backs up, you've got a clog in the waste stack or main line. *See page 57* for instructions on how to clear this type of clog, or contact a drain-cleaning professional.

ANATOMY OF A BATHROOM SINK

Bathroom sinks differ from other sinks (such as kitchen and utility sinks) in that they have a built-in system to handle overflows. Slots or holes near the top of the basin allow water that reaches this preset level to flow through an overflow passage into the drain. Another unique feature of bathroom sinks is the drain stoppers are usually connected to a pop-up mechanism that allows the stopper to be raised or lowered remotely. While handy, this unfortunately creates an area in the drainpipe (where the linkage connects to the stopper) that is prone to clogging.

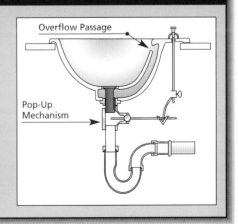

Overflow Passage

Pop-Up Mechanism

Unclogging a Kitchen Sink

1 **Seal off the basin** If your kitchen sink has two basins, you'll need to seal off one side before you can plunge. If you don't, the water the plunger forces down the drain will come up into the other basin instead of working to break the clog free. The simplest way to seal it is to use the black rubber stopper or strainer. In order to use both hands on the plunger, you'll need a helper to hold the strainer in place.

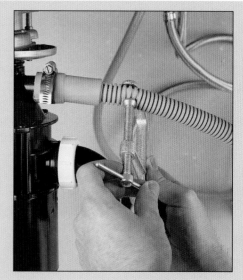

2 **Pinch off the disposer line** If your sink has a garbage disposer built in, you'll need to seal off the hose going to the air gap if there is one. Here again, if you don't seal it off, the water you force down the drain will have an alternate, easier path to follow instead of forcing the clog down the drain. Use a small C-clamp or a pair of vise grips to temporarily pinch the walls of the tubing together.

3 **Fill with water and plunge** Fill the basin with water until it's over the rubber cup of the plunger. I've found that a plunger with a flexible skirt works best here since the skirt fits down into the drain opening, creating a better seal. Give the plunger a series of hard downward strokes and lift it sharply out. Repeat 10 times or so before giving up.

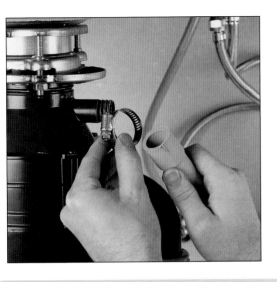

4 **Remove disposer waste line** If one basin works fine but the other is clogged, the problem may be the waste line that connects the disposer to the drain. Place a bucket under the line and loosen the slip nuts that hold it in place. Remove the line and clear any debris that's caught in the pipe.

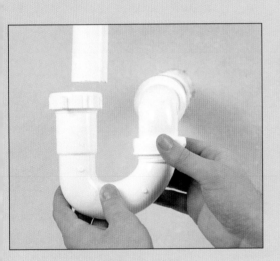

5 **Remove the trap fittings** If both basins are clogged and plunging doesn't clear out the obstruction, the next step is to remove the trap fittings and clear out any debris. Start by placing a bucket under the trap, and have some shop towels handy. Loosen the slip nuts that hold the trap in place, and remove it.

6 **Clear debris or auger** Clear out any debris with a screwdriver or the handle of a wooden spoon. If there isn't any obstruction, the clog is either in the drain line to the sanitary T or in the waste stack itself. In either case, the solution is to clear the clog with a hand auger. *See pages 51–53* for step-by-step instructions on how to use a hand auger. Once the clog is cleared, reassemble the drain system and flush it with water.

Clearing Branch and Main Lines

If plunging and augering the drain line doesn't clear the clog, it's located in either a branch or main waste line, or in the sewer service line. Branch and main line clogs can be cleared by the homeowner by systematically working from the drain line toward the service sewer line. If you're lucky, your home will have clean-out plugs at the end of each branch line. This is the first place to start. If you still don't locate the clog, advance to the main waste stack. For tips on handling a big clog like this, *see the sidebar on the opposite page.* If your problem is in the sewer service line, I suggest calling in a professional drain-service company.

1 **Remove the branch plug** If you are fortunate enough to have clean-out plugs in the ends of your branch lines, locate the one nearest the clogged sink (they're often located between the floor joists in a basement). Before loosening the plug, place a large bucket underneath the opening and have plenty of towels or rags on hand. Don't stand directly beneath the opening, as the backed up water can shoot out of the opening like a geyser. It can surprise you and shoot out a distance before arcing down and eventually trickling to a stop.

2 **Remove the clean-out plug** If you haven't found the clog in one of the branch lines, odds are it's in the main line. You'll need to find the main clean-out and remove the plug. Here again, have on hand a bucket, shops rags, and mop in case the water is backed up into the waste line. A pair of channel-type pliers is all you'll usually need to remove the clean-out plug.

3 **Auger** If you removed the clean-out plug and there was very little water, the clog is likely between the opening and the sanitary T closest to the sink. Here's another instance where a hand auger can come to the rescue. Feed the auger cable in until you feel resistance. Lock it and crank the arm. Continue until the clog is cleared. (*See pages 51–53* for step-by-step directions on using a hand auger.)

DEALING WITH MAJOR CLOGS

Plunging didn't work. Augering the drain and branch lines didn't work. You've got a major clog either in the main waste stack or in the sewer service line.

The simplest way to figure which place is to look in your main drain clean-out. With buckets and towels on hand, carefully unscrew the clean-out plug. If water starts squirting out, the sewer service line is clogged.

You can try to auger the line with a snake, but this is usually best left to a professional drain-service company. If you removed the plug and no water was backed up, the clog is in the waste stack.

If you're feeling adventurous and have access to a long snake, you can try clearing the stack by augering down through the vent stack on your roof. As always, use caution when working with ladders and on a roof.

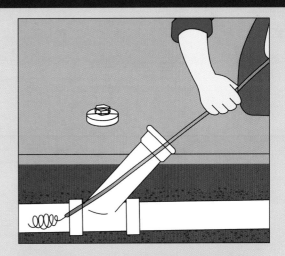

Clearing Shower and Tub Drains

1 **Remove the stopper or strainer** Clumps of hair and soap sludge are the most common culprits for shower and bathtub clogs. The first thing to do is remove the stopper and/or strainer. In many cases, a hair ball can form around the base of the stopper or the strainer. Once it's removed, look down into the drain to see if you see an obstruction. If so, use needlenose pliers to remove it.

2 **For a tub, remove the overflow** If you're working on a bathtub, remove the trip-lever waste and overflow. Do this by removing the two screws that hold the plate near the head of the tub. Then pull out the linkage. Quite often you'll discover a hair ball wrapped around the linkage. If the water flows down the drain when you pull it out, you're in luck—you've found the clog.

3 **Fill with water and plunge** If removing the waste and overflow linkage didn't clear the clog, it's time to pull out the plunger. As with sinks, make sure there's sufficient water in the tub or shower to completely cover the rubber cup. Block off the overflow tube with a damp rag, and plunge with a series of sharp, hard strokes. Continue until the drain is cleared.

Unclogging Toilets

Clogs in toilets occur primarily in the internal trap and are caused by fecal matter, facial tissue, paper towels, cloth diapers, and sanitary napkins. If you know the clog is a cloth diaper or a sanitary napkin, your best bet is to snag it with a closet auger and pull it out. Don't try to force it down the drain with a plunger—it'll most likely get stuck in the main waste vent, and you'll have a bigger problem.

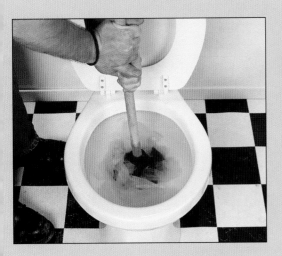

Clear with a plunger A plunger with a skirt or flange works best for plunging toilets, since it creates a better seal. Here again, you'll need sufficient water to cover the rubber cup before plunging. Plunge with vigor here. You'll likely splash a bit, so have some towels handy and wrap one around the base of the toilet. Rest between spurts of plunging, and maintain the water level well over the cup of the plunger.

Or clear with a closet auger If plunging doesn't get the job done, try a closet auger. Turn the handle of the auger as you insert the end into the trap opening. As you press the end into the opening, continue to turn the handle. A series of turns will most likely clear an obstruction. The other option is to try and snag the obstruction (like a cloth diaper) and pull it back out.

Repairing Minor Pipe Leaks

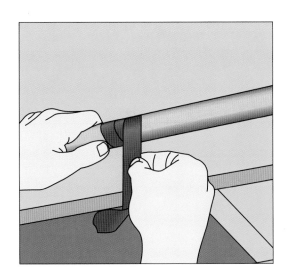

A quick tape fix One of the simplest ways to temporarily stop a small leak is to wrap a generous layer of stout tape (such as electrician's tape or duct tape) around the pipe. (Note: Make sure to turn off the water supply to the pipe and dry it thoroughly before applying tape the pipe.) Keep in mind that this is only a temporary fix until the leak can be properly repaired.

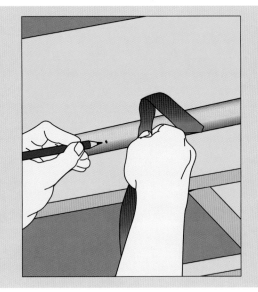

Pencil lead and tape For slightly larger leaks, you can temporarily stymie the flow of water by inserting the tip of a pencil into the hole and snapping it off. To keep it in place, wrap a couple layers of electrician's tape or duct tape around the hole. Here again, turn off the water first and dry the pipe.

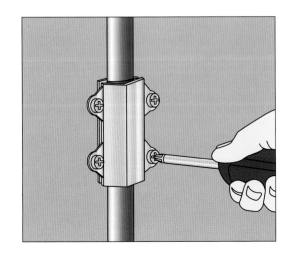

Rubber scraps and hose clamp Another option that's more permanent is to wrap the pipe with scraps of rubber held in place with hose clamps. I always suggest to home-owners that they keep a pipe-repair kit (like the one *shown here*) on hand. These are available at any hardware store and are just two squares of flexible rubber that are squeezed tightly against the pipe with a pair of metal clamps.

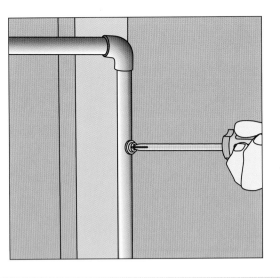

Screw and faucet washer If you've got a small sheet metal screw and a faucet washer rolling around in your junk drawer, they can be pressed into service to temporarily stop a small leak in copper pipe. Use a screwdriver to thread the screw through the faucet washer and then into the hole; just make sure the screw is short enough so it won't puncture the opposite side of the pipe.

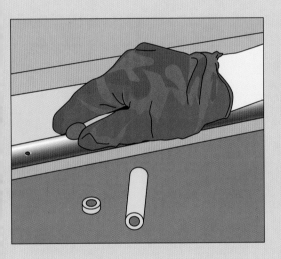

Epoxy putty You'll find epoxy putty at virtually every plumbing supply house in the world—and for good reason: It's easy to use and it's waterproof. Most versions come in a Tootsie Roll–like log that consists of two different color inner and outer layers. When a piece is cut off and the two parts are massaged together, the epoxy is activated (wear gloves for this). Then just apply it to the leak and let it set up before turning the water back on.

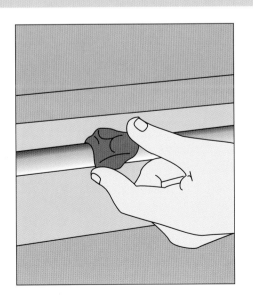

Plumber's putty If you don't have any epoxy putty on hand, a glob of plumber's putty will temporarily staunch the flow from a small leak. Again, this is just a temporary fix until the pipe can be properly repaired. Pull a generous chuck of putty out of the can and roll it around in your hand until it's flexible. Then apply it to the pipe.

Repairing Major Pipe Leaks

Many of the major pipe leaks I've encountered have been caused by water freezing in the pipe. In some cases, the ice will expand to the point of causing the pipe to burst. Or the ice will break a sweated joint. Regardless of the type of damage, the first step is to turn off the water and thaw the frozen pipe. Once you've gotten all the ice and water out of the pipe, it can be repaired. If the pipe is burst, you can patch it with a pipe-repair kit, or replace the damaged section with new pipe. For joints where the ice caused the joint to fail, you can usually just reheat the joint and apply solder to create a new bond; *see pages 38–39* for more on how to do this.

1 **Thaw a frozen pipe** There are a number of ways to thaw a frozen pipe. You can use a heat gun (or hair dryer) or a propane torch (*as shown*), or wrap the pipe with rags and pour hot water on it. Once the pipe has thawed, you'll want to take corrective measures to heat the space where the pipe is located to prevent it from freezing up again in the future.

2 **Smooth rough edges** If you're planning on using a pipe-repair kit, the first thing you'll want to do is remove any exposed rough metal edges. A small metal mill file will make quick work of this. The object here is to file the edges off flush with the outside of the pipe.

3 **Install rubber repair sleeve** With the rough edges removed, you can apply the rubber repair sleeve(s) to the pipe. Some kits supply one large piece that is wrapped around the pipe and cut to fit; others provide two pieces that are sandwiched between the metal clamp plates.

4 **Attach metal clamps** Holding the rubber sleeve(s) in place, position the metal clamps loosely around the pipe. Thread the screws in by hand until they're snug. Then tighten the nuts with an adjustable wrench, each in turn, until the clamp is held securely in place. Turn the water supply back on slowly, and check for leaks.

5 **Or splice in a new section** For a more permanent repair, use a tubing cutter to remove the damaged section. Cut a new piece to fit, and attach it to the existing pipe with a pair of unions. After you've checked the fit, clean the pipe and fittings, apply flux, and sweat the joints (*see pages 36–39* for more on this). Turn the water supply on and check for leaks.

Chapter 5
Repairing Faucets

As with many other modern conveniences, most of us don't pay attention to a faucet until it stops working. Fortunately, faucets are simple, and all work on the same premise—when the handle is activated, a valve inside the faucet turns the water on or off. The secret to successfully repairing a faucet is to know what type of valve it uses to control the flow of water.

Regardless of whether the faucet is located in the kitchen or in the bathroom, it's going to be one of five major types: ball, cartridge, compression, diaphragm, or disk. If it's a kitchen faucet with a single handle, it's most likely a ball-type or cartridge-type. You can tell which by activating the handle: If it rotates, it's a ball-type (*page 65*); if it goes up and down, it's a cartridge-type (*page 68*).

For faucets with two handles, it gets a little murky. If the handle rises and falls as you turn it, it's likely one of three types of compression faucets: modern (*page 70*), outdated (*page 72*), or reverse-compression (*page 78*). Or it may be a disk-type (*page 76*) or diaphragm (*page 74*).

Sometimes the only way to tell is to turn off the water, disassemble the faucet, and compare it to the anatomy drawings in this chapter. If you're still stumped, take the faucet to your local plumbing supply house for help.

Once you've identified the type of faucet, you'll be able to repair it. The first step in repairing it is to identify where the water is coming from. If it's dripping from the spout, it's the valve itself (a stem washer, disk, or diaphragm). For water leaking out from around the handle, it's the seal (either an O-ring or packing washer). When water leaks out from under a base plate, splashed water is working underneath it and then seeping out; *see page 80* for the fix.

The second step is finding the correct parts and replacing them—check your local hardware store for repair kits, or go to a nearby plumbing supply house.

Ball-Type Faucets

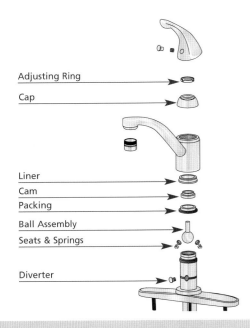

Adjusting Ring

Cap

Liner
Cam
Packing

Ball Assembly
Seats & Springs

Diverter

Anatomy The heart of a ball-type faucet is a ball that rotates as the single handle is pivoted. As the ball rotates, slots in the bottom align with the hot and cold water ports in the faucet body. A pair of spring-loaded seats press up against the ball to serve as simple on-off valves. When the slots align with the seat, the spring pushes up the seat, allowing water to flow out of the spout.

1 Remove the handle If you experience leaks around the handle, the first thing to try is to tighten the adjusting ring *(see Step 2)*. In order to do this, you'll first need to remove the handle. In most cases, it's held in place with a setscrew near its base. Use an Allen wrench to loosen it, and lift the handle off.

2 Tighten the adjusting ring Not all ball-type faucets have an adjusting ring. If yours doesn't, or if tightening the adjusting ring doesn't stop the leak, the problem most likely is either a worn O-ring *(see Step 8)* or worn seats *(see Step 6)*. Most faucet manufacturers include a special tool in their faucet-repair kits for tightening the adjusting ring. Simply insert the tool in the ring and twist clockwise to tighten.

3 **Remove the cap** Start by turning off water to the faucet via the shutoff valves. To access the internal parts of a ball-type faucet, remove the cap with an adjustable wrench (if there are flats on the sides of the cap) or with a pair of channel-type pliers. Since most faucet caps are chromed, it's a good idea to wrap the serrated jaws of the pliers with a turn or two of duct tape to prevent the jaws from scratching the finish.

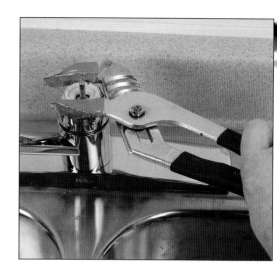

4 **Remove the cam assembly** With the cap off, you can remove the cam assembly. The underside of the cam assembly is concave, to accept the ball and hold it firmly in place yet still allow it to pivot freely. You should be able to lift out the cam assembly by hand; but on a stubborn one, you may need to gently pry it out with a small screwdriver.

5 **Remove the ball** To access the seats that control the water flow, lift out the rotating ball. Be sure to make a note of how it's positioned in the faucet body so that you can reassemble it properly later. If the ball is obviously scratched, bent, or damaged, replace it. You can purchase a plastic or metal replacement ball—the metal varieties cost a bit more but generally last longer.

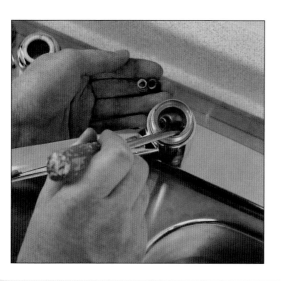

6 **Remove the seats** With a screwdriver or a pair of long-nosed pliers, pry out the seats and springs in the base of the faucet body. Replace all of these with parts provided in the faucet-repair kit, or purchase them separately from a plumbing supply house. Make sure they're seated properly before you reassemble the faucet.

7 **Remove the spout** If you're rebuilding the faucet, or if replacing the seats didn't stop the leak, the spout O-ring may need to be replaced *(see Step 8)*. To get to the O-ring, you'll first need to remove the spout. Grasp the spout firmly and twist it while pulling up. If the O-rings appear to be in good shape, try lubricating them with petroleum jelly and reassemble.

8 **Install new O-rings** If the spout O-rings appear worn, pry them off with a small screwdriver. Then lubricate the new rings with petroleum jelly and slip them over the faucet body. Position the spout over the body and press it down gently while rotating until it seats against the base. Reassemble the faucet, turn on the water, and test.

Cartridge-Type Faucets

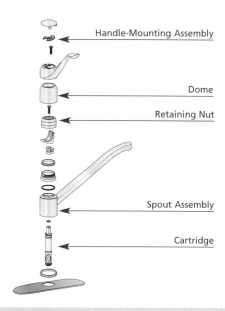

Handle-Mounting Assembly

Dome

Retaining Nut

Spout Assembly

Cartridge

Anatomy In many ways, the cartridge-type faucet works just like a rotating-ball faucet—as the single handle is pivoted, varying amounts of hot and cold water are directed up into the spout. The difference is how the water is directed. Instead of a ball that rotates against spring-loaded seats, a cartridge moves up and down within the faucet body to control the amount and temperature of the water directed to the spout.

1 **Remove the cap and handle** Before you begin work, close the water shutoff valves for the faucet. Use a small screwdriver to pry off the trim cap that conceals the handle screw. Then remove the screw and lift the handle off the faucet body. On some faucets, the handle is held in place by a lip on the retaining nut; tilt the handle to disengage the lip and lift it off.

2 **Remove the handle-mounting assembly** The handles on some cartridge-type faucets attach to the cartridge by way of a mounting assembly. To get to the cartridge, you'll have to remove it. It's typically held in place with a single screw that can be easily removed with a screwdriver. With the screw removed, lift off the mounting assembly.

3 **Remove the retaining nut** Remove the retaining nut with an adjustable wrench (if there are flats on the side) or with a pair of channel-type pliers. Wrap the serrated jaws of the pliers with duct tape to prevent scratching the finish. Lift out the nut and faucet body.

4 **Remove the retainer clip** The cartridge is typically held in place with a U-shaped retainer clip. Use a pair of long-nosed pliers or a screwdriver to pry out the clip from its slot and remove it. Set the clip aside for assembly later—whenever I take apart something with small parts, I set them in a small plastic cup (like a yogurt container) or in an egg carton, with its individual compartments.

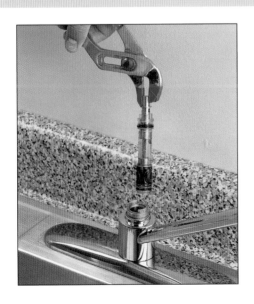

5 **Pull out the cartridge** Wrap tape around pliers to prevent scratching the stem, grab it firmly, and lift it out. If the stem is worn or damaged, replace it—just make sure to align its "ears" properly with the faucet body. If you're rebuilding the faucet, the kit will include new spout O-rings. Lift the spout off the faucet body and pry off the old rings. Lubricate the new O-rings with petroleum jelly and roll them into the correct grooves. Set the spout back in place and reassemble the faucet. Turn on the water and test. If hot and cold are reversed, turn off the water, disassemble, and rotate the stem one half turn.

Modern Compression Faucets

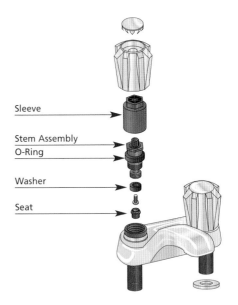

Sleeve

Stem Assembly

O-Ring

Washer

Seat

Anatomy Newer compression faucets (often called seat-and-washer faucets) individually control the flow of hot water and cold water that is then sent to the spout to mix. As the handle is rotated, a valve stem raises or lowers to allow more or less water through to the spout. A rubber washer on the end of the stem presses against the seat in the base of the faucet body to stop the flow when the handle is turned off.

1 Remove the handle and stem If water drips out from the spout when the handles are turned off, the problem is most likely a washer. If it leaks from the handle, it's the O-rings. In either case, start by turning off the water at the shutoff valves. Then remove the handle screws and lift off the handles. Use a pair of channel-type pliers or an adjustable wrench to loosen the locknut holding the stem in place; lift it out of the faucet body.

2 Remove the washer screw and washer To replace the stem washer, hold the stem in one hand and use a screwdriver to loosen the washer-retaining screw. In some cases, it may help to reinstall the handle on the stem for better leverage. If the screw is stubborn, apply a few drops of penetrating oil and let it sit for 15 minutes before trying again. Remove the screw and pry out the old washer.

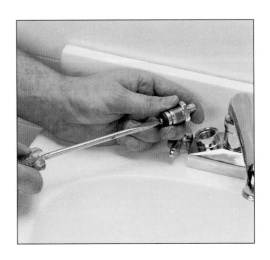

3 **Unscrew the spindle** The stem should move up and down smoothly inside the spindle. If it doesn't, unscrew the stem completely from the spindle. Apply some petroleum jelly or, better yet, "key grease" (found at plumbing supply houses). Key grease is tasteless and odorless, and it can withstand high temperatures. Reassemble the stem and spindle and test for smooth operation. If it's still stiff, replace it.

4 **Replace the O-ring** To stop leaks around the handles, replace the O-ring on the stem. Pry off the old O-ring with a small screwdriver. Lubricate the new O-ring with key grease or petroleum jelly, and roll it into the appropriate groove.

5 **Replace washer and reassemble** Replace the washer with a direct replacement part—you can find most types at any hardware store or at a plumbing supply house. Position the new washer and install the retaining screw. Reassemble the faucet, turn the water back on, and test. If the faucet still drips with the handle off, you may need to replace the seats; *see page 73.*

Outdated Compression Faucets

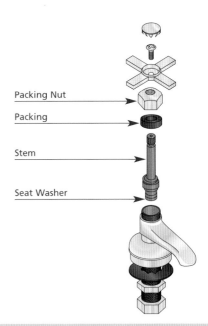

Packing Nut

Packing

Stem

Seat Washer

Anatomy Older compression faucets work much like their modern cousins—the stem and washer systems are almost identical. The major difference is how the faucet prevents water from leaking around the handle. Older versions use a packing washer instead of the more modern O-rings. Sometimes, all it takes to stop a leak is to tighten the packing nut. If that doesn't take care of it, you'll need to replace the packing.

1 **Remove the old packing** To remove the old packing, first turn off the water supply to the faucet. Then pry off the trim cap with a small screwdriver and remove the handle screw. Lift off the handle and set it aside. For stubborn handles, use a faucet-handle puller—often you can borrow one of these from your friendly plumbing supply house. Then pry out the packing from the packing nut.

2 **Change the packing** You can still find packing washers at many hardware stores—if you can't, try a plumbing supply house—or wrap a couple strands of graphite-impregnated string (available at most hardware stores) around the stem. Thread the packing nut back on, hand-tighten it, and then tighten it one quarter turn past that. Turn on the water and test. If it still leaks, tighten another quarter turn. Repeat until the leak stops.

Valve Seats

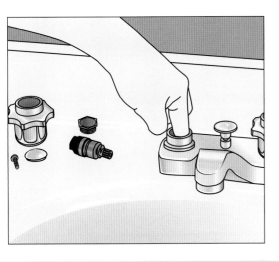

1 **Check for damage** If replacing a seat washer does not stop a faucet from dripping, the problem is likely a damaged seat. To check for damage, turn off the water and remove the handle and stem. Insert a finger into the faucet body, and gently rub your fingertip around the lip of the seat—it should be smooth and level. If you find burrs, scratches, or other damage, replace the seat with an identical replacement part.

2 **Remove the seat** Although it is possible to remove some seats with an Allen wrench, I've found that it's worth the money to purchase a seat wrench. This special L-shaped tool is specifically designed for the job. It has a different tip on each end to fit just about any seat; and the larger L-shaped handle provides the leverage you'll need to loosen it. Stubborn seats often require penetrating oil to convince them to loosen up.

3 **Install a new seat** Once you've removed the old seat, you can install the new one. Most plumbers will apply a couple turns of Teflon tape around the threads of the seat before installing it. The easiest way I've found to install the new seat is to slip it onto the end of the seat wrench and then slide this into the faucet body. Tighten the seat with the seat wrench one quarter turn past hand-tight. Then reassemble the faucet, turn on the water, and test.

Diaphragm Faucets

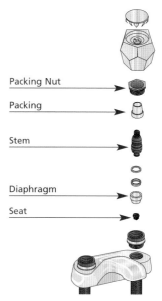

Anatomy With its stem and seat, the diaphragm faucet looks and works much like any other compression faucet. What's different is it doesn't use a seat washer. Instead it uses a much more reliable diaphragm—the edges of the rubber diaphragm wrap around the stem and overlap at the edges to create an excellent seal. When the handle is closed, the bottom of the diaphragm presses against the seat, stopping the flow of water.

Packing Nut

Packing

Stem

Diaphragm

Seat

1 **Remove the cap and handle** To replace the diaphragm, first turn off the water at the shutoff valves. Then use a small screwdriver to pry off the trim cap on the handle. Remove the handle screw and lift off the handle.

2 **Remove the locknut** The stem on most diaphragm faucets is held in place with a stem locknut. Use an adjustable wrench or channel-type pliers (with the jaws taped to prevent damage) to loosen the nut. Once loose, unthread it completely, lift it off, and set it aside.

3 **Remove the stem** With the stem locknut removed, pull the stem up and out of the faucet body. Examine the stem for wear and damage, checking the threads closely for burrs, which can cause rough operation. If necessary, purchase an exact replacement part from your local plumbing supply house.

4 **Replace the diaphragm** If the diaphragm itself is worn, remove it by peeling it off of the stem. For proper operation, it's imperative that you replace it with an identical part, not one that just looks like a close fit—it'll leak for sure. Press the new diaphragm on the end of the stem, and rotate it a half turn or so to make sure the rubber edges overlap the stem properly.

5 **Reassemble** Apply a thin coat of key grease (*see Step 3 on page 71*) to the stem and reinsert it into the faucet body. Thread on the stem locknut so it's hand-tight. Then tighten it one quarter turn and install the handle. Turn on the water and check for leaks. If necessary, remove the handle and give the stem locknut another quarter turn. Repeat as necessary to eliminate any leaks.

Disk Faucets

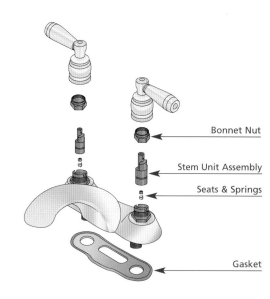

Anatomy Disk faucets are a hybrid of a rotating ball and a cartridge faucet. Like the rotating ball, it uses spring-loaded seats. But instead of a ball, it features a cartridge-like stem unit (the disk assembly) that moves up and down with the handle to allow water to flow. A defective disk assembly can cause water to leak from the handle or the spout. Often, all it takes to remedy this is cleaning the disk assembly or replacing the seals.

Bonnet Nut

Stem Unit Assembly

Seats & Springs

Gasket

❶ Remove the handle(s) Turn off the water supply at the shutoff valves and remove the handle(s). Disk faucets are available in one- or two-handled versions. For a one-handled faucet, use an Allen wrench to loosen the setscrew that secures the handle; lift it off and set it aside. For a two-handled faucet, pry off the trim caps with a screwdriver; remove the handle screws and lift off the handles.

❷ Lift out the disk assembly In order to remove the disk assembly, you'll first need to unscrew the escutcheon cap (on a one-handled faucet) or bonnet nuts (for two-handled faucets). In either case, use a pair of channel-type pliers (with the jaws taped to prevent scratching the finish) to loosen the cap or nut. Remove the cap or nut and then lift out the disk assembly.

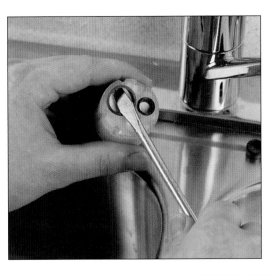

3 **Remove old seals** The neoprene seals are located on the bottom of the disk assembly. Single-handled assemblies have one neoprene seal; double-handled faucets have two. Pry the seals out with the tip of a screwdriver, taking care not to scratch the disk assembly. If the disk assembly is cracked or damaged, it will need to be replaced; *see Step 5.* Otherwise, proceed to the next step.

4 **Clean the opening and replace the seals** In some areas of the country, deposits can build up on the base of the disk assembly due to the local water. If this happens, use an abrasive pad (such as Scotch-Brite) to scour away the deposit. Soaking the assembly in vinegar can also help remove stubborn deposits. Pay particular attention to the area around the openings, as buildups here can cause the biggest problems. If the seals are worn or damaged, replace them.

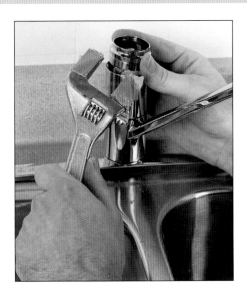

5 **Replace the cylinder if necessary** Reassemble the faucet by inserting the disk assembly in the faucet body—make sure to align the seals in the assembly with the water inlets. Thread on the cap or bonnet nuts and slip on the handle. Slowly turn on the water and check for leaks. If the faucet continues to leak after cleaning, the cylinder needs to be replaced—make sure to get an identical replacement part.

Reverse-Compression Faucets

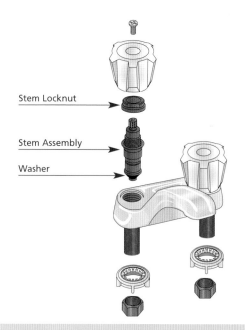

Stem Locknut

Stem Assembly

Washer

Anatomy A reverse-compression faucet works in reverse compared to a standard compression faucet. As the handle is turned on, the spindle moves downward, creating a gap between the washer and the seat, allowing water to flow. On a compression faucet, the spindle moves up to let water in. Internally, the parts are similar, and it's not until you watch the spindle react to turning the handle that you'll know what variety you've got: reverse or standard.

1 **Remove the cap, handle, and nut** Just as with a compression faucet, if you've got a leak, try tightening the packing nut before disassembling anything. If this doesn't work, turn off the water supply at the shutoff valves; then pry off the trim cap with a small screwdriver, remove the handle screw, and lift off the handle. Next, loosen the packing nut with an adjustable wrench. If the nut is chromed, tape the jaws of the pliers to prevent scratching the finish.

2 **Lift out the stem** With the packing nut removed, you'll be able to lift out the stem. If the stem is cracked or worn, replace it. If it's in good shape, go to the next step. Some stem assemblies will use an O-ring instead of a packing washer. If this is the case, inspect the O-ring for wear. If in doubt, replace it. Apply a thin coat of petroleum jelly onto the new O-ring and roll it into the correct groove.

3 **Replace the packing washer** If your faucet leaks and it uses a packing washer, replace it with an exact replacement part. If you can't find one at the local hardware store, try a plumbing supply house. If they don't have it, wrap a couple strands of graphite-impregnated string around the stem.

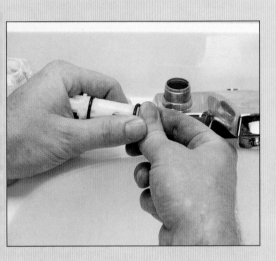

4 **Replace the seat washer** Because the seats on reverse-compression faucets don't get as much abuse as on a standard faucet, the seat washers tend to last a long time. But in the event the washer has seen better times, it's easy to replace. Just remove the seat washer screw and install an exact replacement. If the screw is stubborn, apply some penetrating oil and wait 15 minutes before trying again.

5 **Reassemble** Insert the stem into the faucet body. Thread on the packing nut until it's hand-tight. Next, use an adjustable wrench to further tighten it one quarter turn. Slip the handle on and slowly turn on the water. If it leaks, remove the handle and tighten the packing nut another quarter turn. Repeat until there are no leaks.

Leaky Base Plates

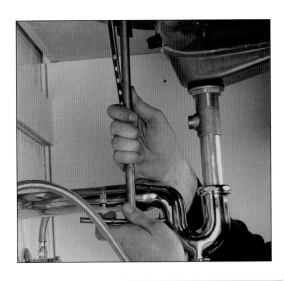

1 **Tighten the mounting nuts** If you have water seeping out from under a base plate, it may be that the base plate is not being pulled down firmly against the sink. To remedy this, tighten the faucet-mounting nuts. Since there's rarely sufficient room to turn a wrench under a sink, use a basin wrench, with its long extension handle, to reach up and tighten the nuts.

2 **Check for gap in base plate** If tightening the mounting nuts didn't solve the problem, it's likely that splashed water is working its way under the cavity in the base plate or the faucet base. The solution is to fill the cavity with plumber's putty. First, try loosening the mounting nuts and see if you can lift the faucet body up enough to force putty in. Use a putty knife, and be generous—any excess will be forced out when you tighten the mounting nuts. Slowly tighten the mounting nuts, alternating from side to side. Scrape off any excess with a plastic putty knife or spatula to prevent scratching the sink or countertop.

3 **Remove the faucet** In situations where you can't force plumber's putty under the base plate or faucet base, you'll have to remove the faucet to apply the putty. Turn off the water supply, turn on the faucet to drain out any remaining water, and disconnect the supply lines with an adjustable wrench. Then remove the mounting nuts with a basin wrench and lift out the faucet.

4 **Remove the old putty** Once you've removed the faucet, take the time to remove any of the old plumber's putty that has stuck to the sink. Scrape off the old putty from both the faucet and the sink; using a plastic putty knife or spatula will prevent scratching the sink.

5 **Pack with plumber's putty** With the old putty removed, apply a generous amount to the base plate or faucet body. Fill the entire cavity to prevent splashed water from working its way into it. Position the faucet on the sink or countertop, and thread the mounting nuts on. Tighten them slowly, alternating form side to side. Scrape off any excess putty with a plastic putty knife or spatula. Reconnect the supply lines, restore the water, and check for leaks.

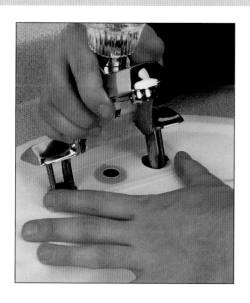

6 **Or install a new gasket** On some faucets, the problem is an aging or damaged gasket that fits between the faucet body and the sink or countertop. Here again, you'll need to remove the faucet before you can replace the gasket; *see Step 3 on the opposite page.* Buy an exact replacement gasket and reassemble the faucet. If you can't find an exact replacement, pack the cavity with plumber's putty and then reassemble.

Chapter 6
Repairing Sinks

A sink installed in a bathroom and one installed in a kitchen have many things in common. Both have a pair of lines running up into the faucet to supply hot and cold water and are controlled by separate shutoff valves. Both have a trap installed on the waste line to block sewer gas from passing up into the house. And both use some sort of stopper to close the drain so the basin(s) can be filled.

Because of this, many of the problems you'll encounter with sinks have similar solutions. Leaks can often be eliminated with some fresh plumber's putty or tightening a slip nut on a drain body or trap, or tightening the clamp on a hose (*see pages 83 and 87*). There are, however, problems unique to each style of sink.

Bathroom sinks with their integral pop-up drain stoppers often need periodic attention to maintain a proper seal for their drain (*page 84*). Also, when you combine this built-in pop-up mechanism with waste lines that are smaller in diameter than those of kitchen sinks, you end up with frequent clogs. (For step-by-step details

on how to deal with bathroom sink clogs, *see page 49.*)

Although kitchen sinks don't have built-in pop-up drain stoppers, their typical double-basin design and their common hookups to garbage disposers (*page 88*) and dishwashers (*page 87*) make them just as susceptible to clogs. (*See page 54* for details on kitchen sink clogs.)

Kitchen sinks also get worked harder than bathroom sinks. The constant one-two punch of hot water and periodic banging of pots and pans plus the vibrations caused by the disposer and dishwasher take their toll on gaskets, seals, and hoses. Fortunately, rubber gaskets and hoses are easy to replace, and many seals can be refreshed with a few cents' worth of plumber's putty (*page 86*). (**Safety Note:** Before you work on any dishwasher or garbage disposer, make sure you turn off the power.)

Stopping Bathroom Sink Leaks

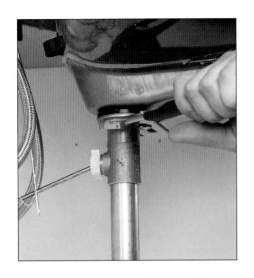

1 **Tighten the lift-rod retaining nut** A common cause of leaks under a bathroom sink is a loose connection between the drain and the linkage connecting the pop-up mechanism. If you detect water seeping out from around the mechanism, tighten the lift-rod retaining nut with a pair of channel-type pliers. If this doesn't eliminate the leak, remove the retaining nut and replace the washer under the nut.

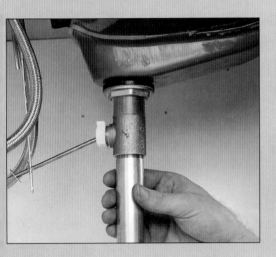

2 **Tighten the tailpiece** Another source of leaks under the sink is a tailpiece that has worked loose—often the result of being bumped by objects placed under the sink. This is a common problem, as the tailpiece joints are most often held together with slip nuts. To stop the leak, tighten the tailpiece by hand until it's snug, and then tighten it another quarter turn. If the leak persists, replace the tailpiece compression ring.

3 **Replace the sink flange** If water is leaking out from around the base of the sink flange, you'll need to either install fresh putty around the flange or, if the flange is cracked, replace it. Start by removing the trap and drain body. Push the flange out of the sink from underneath and remove the old putty. If the flange looks good, apply a coil of plumber's putty to the flange lip and reinstall it. Replace a broken or deformed flange.

Adjusting a Pop-Up Mechanism

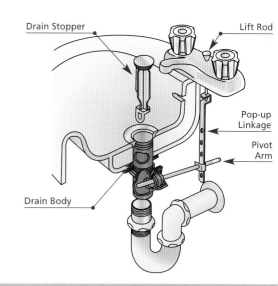

Drain Stopper

Lift Rod

Pop-up Linkage

Pivot Arm

Drain Body

Anatomy A pop-up mechanism consists of three main parts: the lift arm, the linkage, and the drain stopper. The lift arm is connected to the pop-up linkage, which controls the pivot arm. The drain stopper rests on (or is hooked up to) the pivot arm. Raising or lowering the handle of the lift arm forces the pivot arm to move up and down, allowing the stopper to open or close the drain.

❶ Adjust the pivot rod If the drain stopper doesn't fully close the drain, the pivot rod may need to be adjusted. Start by pinching the spring clip that connects the pivot rod to the lift arm. Pull the lift arm off the pivot rod, and insert the rod in a different hole in the lift arm. A little trial-and-error testing will be needed here to determine which hole in the lift arm works best.

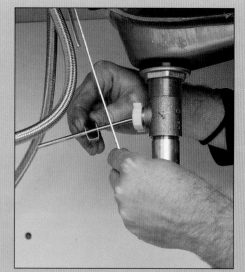

❷ Adjust the lift rod If after you've adjusted the pivot rod the stopper still doesn't seal properly, adjust the lift rod. The lift arm usually consists of two parts: a lift rod and an adjustable linkage that connects to the pivot arm. Loosen the adjustment screw (typically a thumbscrew) on the linkage, and push the linkage up farther on the link rod. Tighten it and test. If it still leaks, you may need to replace the drain body.

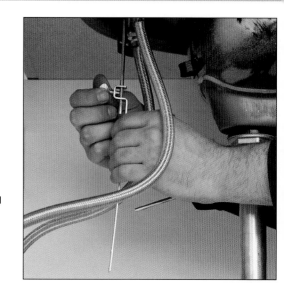

Replacing a Bathroom Sink Drain Body

1 **Remove the old drain body** To replace the drain body on a bathroom sink, start by removing the trap and pop-up mechanism. Using a pair of channel-type pliers, loosen the nut that secures the drain body to the sink. Push the drain body up from underneath, and pry out or unscrew the sink flange.

2 **Install the new drain body** Scrape off any old plumber's putty from around the lip of the drain hole. Apply a coil of fresh putty and insert the new drain body (you'll need to temporarily remove the pivot rod). Slip on any gaskets provided with the new body, and thread on the locknut. Reinstall the pivot arm and rotate the body to align the pivot rod with the lift-arm linkage. Tighten the nut, and remove any excess putty with a plastic putty knife or spatula.

Types of Drain Bodies Drain bodies are available in either one-piece or two-piece models and may be made of plastic or brass. The bodies on the two-piece versions are threaded and simply screw together. Although plastic drain bodies are easy to find at almost any home center, it's worth the extra effort to hunt down a brass one—the plastic versions just don't hold up very well over time.

Fixing a Leaky Kitchen Strainer

1 **Remove the strainer** Because of the constant abuse it receives, it's just a matter of time before the strainer on a kitchen sink begins to leak. In most cases, this occurs because the putty under the flange has dried out. If the strainer is cracked or split, it needs to be replaced. In either case, start by removing the tailpiece, then use a spud wrench or large channel-type pliers to loosen the retaining nut and lift out the strainer.

2 **Reinstall with fresh putty** Scrape off any old plumber's putty from around the lip. Do the same for the old strainer if you're going to reuse it. Now apply a fresh coil of putty around the lip of the strainer. Insert the body of the strainer into the opening, and slip on any rubber gaskets from underneath. Thread on the retaining nut and tighten it.

3 **Clean off excess putty** Tightening the retaining nut will force out any excess plumber's putty from under the flange of the strainer body. Remove this with a putty knife, being careful not to scratch the sink. (A no-scratch alternative is to use a plastic putty knife.)

Repairing Dishwasher Connections

1 **Remove the drain hose** Leaks under a kitchen sink are sometimes caused by the connections to the dishwasher. The hose that runs between the dishwasher and the disposer is a common culprit. Since it transports hot water, it can dry out and split. If it has split, turn off power to the dishwasher and disposer. Loosen the hose clamps the remove the hose. If just the ends are spilt, go to Step 2. Otherwise, replace the hose.

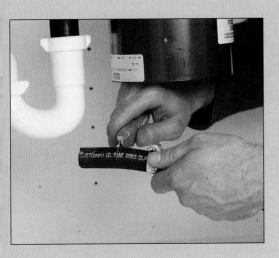

2 **Trim the hose** If the hose is long enough, you can refurbish it by simply trimming off the ends that are cracked or split. Use a sharp utility knife to pare away any damage. Then slip on the hose clamps and reconnect. Turn power back on and test. If the hose still leaks, replace it.

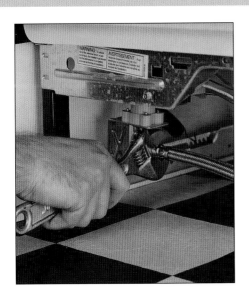

3 **Check the dishwasher connection** Water leaking out from under a sink may actually be coming from a nearby dishwasher. To check this, turn off power to the dishwasher, and remove the service panel at the front bottom of the dishwasher (it may be held in place with spring clips or screws). The leak is probably caused by a loose connection to the water inlet line. Tighten it with an adjustable wrench, replace the service panel, and test.

Stopping Garbage Disposer Leaks

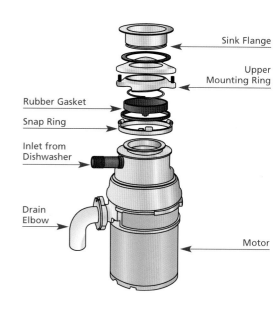

Sink Flange

Upper Mounting Ring

Rubber Gasket

Snap Ring

Inlet from Dishwasher

Drain Elbow

Motor

Anatomy of a disposer Most garbage disposers connect to the sink via a mounting bracket cushioned by a large rubber gasket and held in place with a snap ring. Since these machines are subjected to constant heat and vibration, the rubber gasket that provides the seal between the sink and the disposer will often degrade and fail.

1 **Disconnect and remove the disposer** To replace the rubber gasket, turn off the power to the disposer (or unplug it). Then slip a bucket under it and remove the disposer drainpipe and the air-gap hose (if there is one). Depending on how your disposer is set up, you may also need to remove the trap. Support the disposer with one hand and loosen the snap ring (you may have to use a screwdriver to persuade it to move).

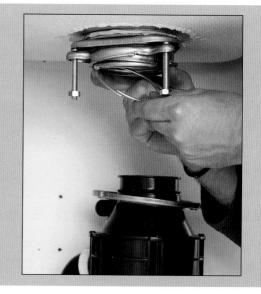

2 **Replace the disposer gasket** Once you've disconnected the disposer, set it down gently in the base of the cabinet, being careful not to pull on its electrical cable if it's hard-wired. Pull off the rubber gasket and replace it with an exact replacement. Your disposer may have additional gaskets that may need to be replaced; consult your owner's manual. Reinstall the disposer by reversing the steps you took to remove it.

Repairing a Diverter Valve

1 **Remove the faucet handle** Low water pressure to a sprayer may be caused by a defective diverter valve inside the faucet body. This valve diverts the water from the faucet to the sprayer when the sprayer is activated. To service the valve, you'll need to remove the handle. Start by closing the water shutoffs to the faucet. Then remove the handle. (*See Chapter 5* for handle-removal instructions for the different faucet types.)

2 **Remove the diverter valve** The diverter valve is usually press-fit into the faucet body. It can be removed by pulling it out with a pair of needlenose pliers. If it's encrusted with deposits, soak it in vinegar for a half-hour or so and then scrub it clean with a toothbrush. If it comes clean, reinstall it and check for proper operation.

3 **Replace worn O-rings** If cleaning the diverter valve doesn't help, try replacing any O-rings before replacing the entire diverter valve. Coat the new O-rings with key grease or petroleum jelly, reinstall the diverter valve, and reassemble the faucet.

Replacing a Spray Hose

1 **Unscrew the sprayer** Another cause of low water pressure to a sprayer is a hose that is defective or constricted due to water deposits. In both cases, the simplest solution is to replace the hose. Start by using an adjustable wrench to loosen the nut that connects the spray hose to the faucet body. Then unscrew it by hand.

2 **Disconnect the spray head** At the other end of the hose, disconnect the spray head by unscrewing it from its handle mount. As long as you've got the spray head off, it's a good time to replace the washer and check the aerator to make sure there are no deposits clogging it up. (*See page 91* for more on aerators.)

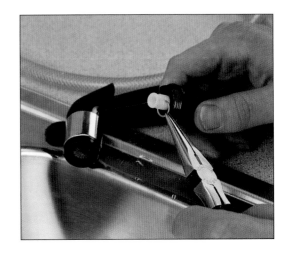

3 **Remove the retaining clip** The handle mount of the spray head is often held in place on the hose by a retaining clip. Remove the clip with a pair of needlenose pliers. Discard the old hose and install the new one, reassembling it in the reverse order that you took it apart.

Fixing an Aerator

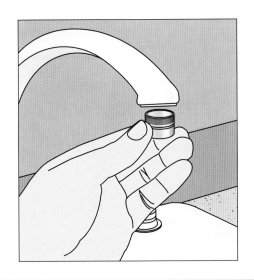

1 **Remove it from the faucet** A clogged aerator not only reduces water flow, but can also cause the water that does come out to spray in a wild or erratic pattern. The solution in most cases is a simple cleaning. Unscrew the aerator by hand to remove it.

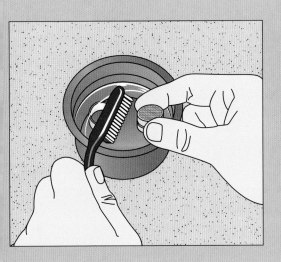

2 **Clean the screen** Soak the aerator in vinegar for a half-hour or so to loosen up any stubborn sediment. Then scrub it vigorously with an old toothbrush. If the sediment is hardened and can't be removed with cleaning, you'll need to buy a replacement. Be especially careful as you thread the aerator back on the spout: The threads are very fine and are easily cross-threaded.

Types of Aerators If it's time to replace your aerator, consider upgrading it to a swivel aerator. Swivel aerators are available in a wide variety of colors, options, and materials. In addition to swiveling, most versions have a pull-down spray feature that converts the faucet into a sprayer. Stick with the all-metal versions—they cost a bit more but will outlast their inexpensive plastic cousins.

Chapter 7
Repairing Toilets

Repairing a toilet may not be one of the most glamorous home improvement jobs, but it will be appreciated by everyone in the house. A toilet that runs constantly not only is annoying, it wastes gallons of water a day—which literally is money down the drain. A leaky toilet, on the other hand, may not be quite as annoying, but if not repaired promptly will result in even more money slipping away—in the form of repairs to damaged floors or walls.

Fortunately, most toilet repairs are simple. In many cases, drips can be eliminated by tightening mounting or connecting bolts (*page 104*). If this doesn't do the job, you'll likely have to replace seals or gaskets to stop the leak—such as the wax ring that fits between the toilet bowl and the closet flange in the floor (*page 105*).

One of the most convenient features of a toilet—it automatically refills after it's been flushed—is also a source of mystery for many homeowners. A Rube Goldberg–type contraption inside called a ballcock is the source of the confusion. Although it looks complicated, a ballcock is just a valve that controls the flow of water into the tank; a float ball or cup connected to the ballcock rises or falls along with the water level in the tank to turn the water off or on. (For more on ballcocks, *see page 100.*)

A malfunctioning or improperly adjusted ballcock is the number-one culprit when it comes to a toilet that won't stop running. Sometimes all it takes is repositioning a lift wire or float arm to adjust the water level (*see pages 94 and 95*). Other times, you'll need to disassemble the ballcock and replace a worn-out seal.

If your existing ballcock is old but of good quality, I strongly recommend repairing it. Many plumbing supply houses will carry the necessary seals, gaskets, or O-rings needed to repair any of the three main types of ballcocks: diaphragm (*page 97*), plunger-valve (*page 98*), and float-cup (*page 99*). If you can't find parts, or if the ballcock is damaged beyond repair, it's a simple job to replace it; *see page 100.*

How a Toilet Works

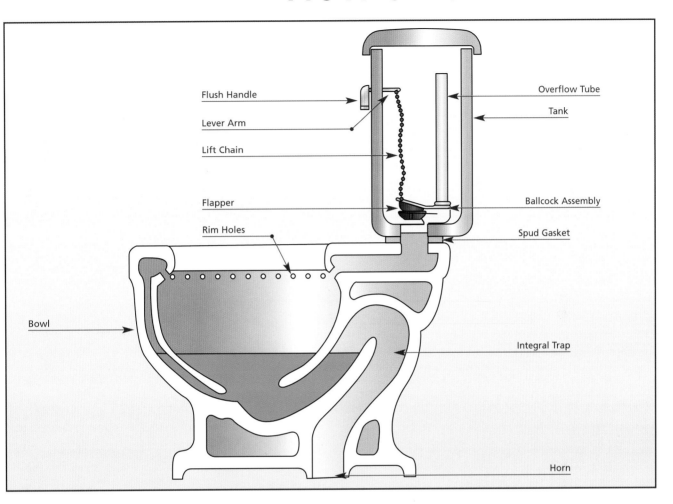

Flush Handle

Lever Arm

Lift Chain

Flapper

Rim Holes

Bowl

Overflow Tube

Tank

Ballcock Assembly

Spud Gasket

Integral Trap

Horn

There a two main parts to a standard toilet: the tank and the bowl. The tank holds a preset amount of water to flush the existing contents of the bowl down the waste line (by code, 1.6 gallons in all toilets installed after 1996, and up to 3.5 gallons in toilets installed prior to 1996). When the flush handle is depressed, the lever arm raises a ball or flapper in the bottom of the tank—by way of either a chain or lift wires—allowing the water in the tank to flow down into the bowl. The water flows through a series of holes in the rim and, with the aid of a little gravity, forces the contents of the bowl to exit through the integral trap, out through the horn, and down into the waste line via a sanitary T. After the majority of the water drains out of the

tank, a flapper or ball will drop down to stop the flow. At the same time, the ballcock assembly inside the tank permits fresh water to flow into the tank, and into the bowl via the overflow tube. When a preset level (controlled by either a float ball, a flow cup, or a metered valve) is reached, the water shuts off. If for any reason the ballcock fails to shut the water off, the water will rise above the overflow tube and drain down into the bowl. A wax ring fitted around the horn sits on the floor and creates a seal between the toilet and the closet flange to prevent both water and sewer gas from leaking out. A spud gasket forms a seal between the tank and the bowl.

Adjusting the Water Level

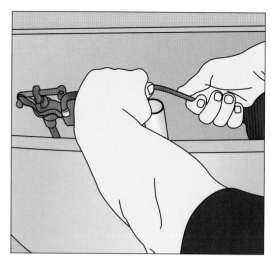

1 **Bend the float arm** If the water level in your toilet tank is too high, it will rise above the overflow tube and drain down into the bowl. If the level is too low, there won't be enough water to clear the bowl. Properly set, the tank water will be ½" to 1" below the top of the overflow tube. For a float arm, lower the water level by bending the float arm down; bend it up to raise the water level.

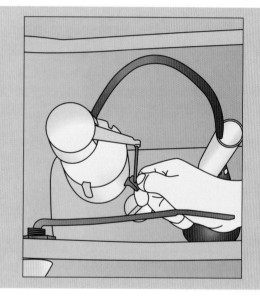

2 **Adjust the float cup** If your toilet uses a float-cup ballcock, you can raise or lower the cup (and consequently the water level) by first pinching the metal retaining clip that fits over a pull rod. Then slide the clip (and cup) up or down to the desired water level and release the clip. Flush the toilet and check the level. Readjust as necessary.

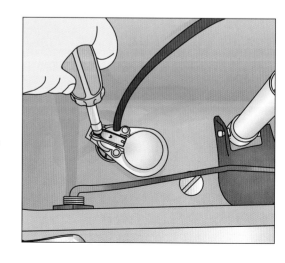

3 **Adjust the fill valve** Some toilet ballcocks use a metered fill valve to control the level of water in the tank. On these models, adjusting the water level is simply a matter of turning an adjustment screw located on top of the ballcock valve. Lower the level by turning the screw counterclockwise; raise it by turning it clockwise. Turn the screw in half-turn increments until the desired level is reached.

Adjusting the Handle and Lift Chain

1 **Adjust the handle** The flush handle on a toilet will occasionally work loose, from constant use. Use an adjustable wrench to tighten the locknut that holds it in place. If the nut won't budge, try some penetrating oil. If it still won't move, cut through the shaft of the handle with a mini-hacksaw and replace the handle assembly.

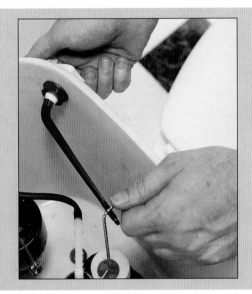

2 **Adjust the lift chain** If a lift chain is too short or is attached too high up on the trip lever, the flapper won't close to form a seal. When the chain is too long or is attached too low on the trip lever, you'll have to hold the handle down to flush the toilet completely. First try moving the chain to another hole in the trip lever. If this doesn't work, remove links from or add links to the chain.

ADJUSTING LIFT WIRES

Some older toilets use lift wires and a tank ball instead of a chain and a flapper to control the flow of waste water in the toilet. There are three parts to this type of flush assembly: two vertical brass wires and a guide (either plastic or metal) that attaches to the overflow tube. The lower lift wire is threaded and screws into the tank ball. It's held vertical by the guide. A hook on the other end accepts the upper lift wire that connects to the trip lever. You can adjust how they work together by bending either wire with a pair of needlenose pliers.

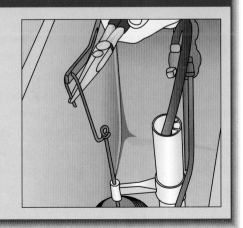

Stopping a Running Toilet

A toilet that runs constantly is a real nuisance. The quickest way to determine why it's running is to remove the tank lid and check the overflow pipe; *see Steps 1 and 2 below.* If this doesn't solve the problem, odds are that there's something wrong with the ballcock. There are three common types of ballcocks: diaphragm, plunger-valve, and float-cup. To repair a diaphragm ballcock, *see page 97.* For plunger-valve and float-cup ballcocks, *see pages 98 and 99, respectively.* For step-by-step directions on replacing a ballcock, *see page 100.* If water isn't flowing into the overflow, the problem is likely caused by a leaky or faulty flush valve. In both cases, water is seeping past the seal between the flapper or ball and the flush valve. First try adjusting and cleaning the flush valve (*see Steps 1 and 2 on page 102*). If this doesn't stop the toilet from running, replace the flapper or ball (*see Step 3 on page 102*). Or replace the flush valve itself (*see page 103*).

1 **Check the overflow pipe** One of the most common causes of a constantly running toilet is water draining over and down into the overflow tube. This can be caused by a couple of things. First, the water level may be set too high in the tank. What you're looking for here is the water to stop ½" to 1" below the top of the overflow tube. If it's too high, *see page 94.*

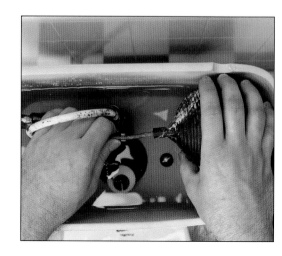

2 **Check the tank ball** If after you've adjusted the water level, water still flows into the overflow tube, the problem may be a defective float ball. Over time, a crack in the ball can allow water to seep in. This added weight can prevent the ball from rising high enough for the ballcock to shut off the water. To check this, unscrew the ball and give a shake. If you hear water sloshing around inside, it's time for a replacement.

Repairing a Diaphragm Ballcock

1 **Remove bonnet screws** Diaphragm ballcocks are made of plastic and use a rubber diaphragm to control the water flow inside the tank. To repair a diaphragm ballcock, first turn off the water to the toilet at the shutoff valve. Then empty the tank by flushing the toilet. To access the rubber diaphragm, use a screwdriver to remove the screws that hold the bonnet on top of the ballcock.

2 **Lift off the float arm** Gently lift off the bonnet and the attached float arm. Inside you'll find the rubber diaphragm and a plastic plunger. Pull the plunger out with your fingers, and gently pry out the diaphragm with a small screwdriver. Both the diaphragm and the O-rings on the plunger should be supple and free from any signs of wear and tear.

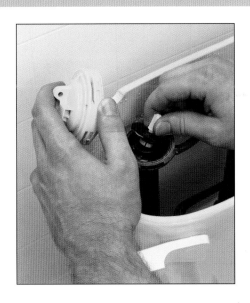

3 **Replace parts as needed** If any of the interior parts are cracked or stiff, replace them with identical replacement parts (see your local plumbing supply house for parts availability). Apply a thin film of petroleum jelly to the O-rings before replacing the plunger. Reassemble the ballcock, turn on water, and check for leaks. If the assembly itself is worn or cracked, you'll need to replace the entire ballcock; *see page 100.*

Repairing a Plunger-Valve Ballcock

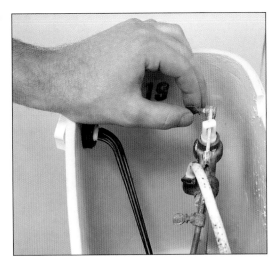

1 **Remove the wing nuts** A traditional plunger-valve ballcock is made of brass, and the water flow is controlled by a plunger that attaches to the float arm. To service a plunger-valve ballcock, first turn off the water supply at the shutoff and empty the tank by flushing the toilet. Then remove the brass wing nuts or thumbscrews on the ballcock. If they're hard to remove, apply penetrating oil, wait 15 minutes, and try again.

2 **Remove the plunger** Now you can pull the plunger out of the ballcock. You'll find a packing washer (this may be a leather washer on older models) or an O-ring that forms a seal to keep water from squirting out of the top of the ballcock. (If water is seeping out, the washer needs to be replaced; *see Step 3.*) On the bottom of the plunger, there's a rubber washer that shuts off the flow of incoming water when the tank is full.

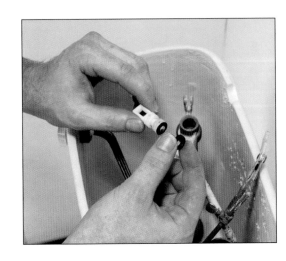

3 **Install replacement washers** If the packing washer or O-ring is worn, pry off the old one and slip on a direct replacement. If the washer on the end of the plunger is worn, remove the stem screw, if it has one (or simply pry it out), and replace the washer. Reassemble the ballcock, turn on the water, and check for leaks.

Repairing a Float-Cup Ballcock

1 **Remove the ballcock cap** Like the diaphragm ballcock, the float-cup ballcock is made of plastic. Water flow is controlled by an arm that connects to a pull rod that fastens onto the float cup. To repair a float-cup ballcock, turn off the water supply at the shutoff and empty the tank by flushing the toilet. Then pull off the plastic cap that covers the ballcock.

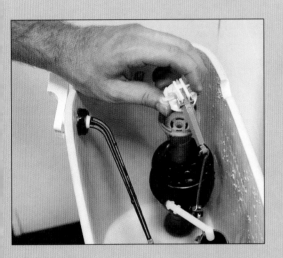

2 **Remove and clean the bonnet** When the cap is removed, you'll see a plastic bonnet inside. This bonnet can be removed by first pressing down on it while twisting counterclockwise. Once unlocked, you can pull out the bonnet. If there is any built-up sediment inside, clean it out with a small wire brush or an old toothbrush.

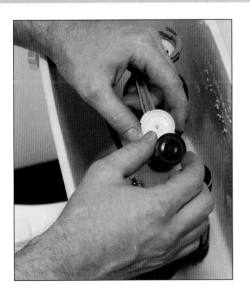

3 **Replace the seal** The only part inside a float-cup ballcock that's really serviceable is the rubber seal. If it's worn or damaged, gently pry it out with a small screwdriver. Replace it with an identical part (see your local plumbing supply house for availability of parts). If the assembly is cracked or worn, replace the entire ballcock; *see page 100.* Reassemble the ballcock, turn on the water, and check for leaks.

Installing a New Ballcock

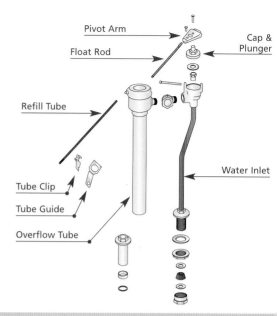

Pivot Arm

Float Rod

Cap & Plunger

Refill Tube

Water Inlet

Tube Clip

Tube Guide

Overflow Tube

Anatomy A ballcock is just a valve—what makes it special is that it's automatic. When a toilet is flushed, tank water flows into the bowl to force its contents down the waste line. As the water level drops, so does the float arm. This in turn pulls up a plunger inside the ballcock, allowing water to flow and fill the tank. The water rises along with the float ball and arm, forcing the plunger down to shut off the water.

1 **Remove the old ballcock** To remove an old ball-cock, start by shutting off the water to the toilet and flushing to empty the tank. Then use a sponge to remove any water remaining in the tank. Use an adjustable wrench to disconnect the coupling nut that runs up from the shutoff valve into the base of the ballcock assembly. Then loosen the mounting nut that holds the ballcock in place. Remove the tube running into the overflow tube, and lift out the old ballcock.

2 **Install the new ballcock** To ensure a good seal when the new ballcock is installed, scrub the area around the hole in the bottom of the tank with an abrasive nylon pad (like Scotch-Brite). Then press the gasket provided with the new ballcock onto its tailpiece. Next, insert the tailpiece of the ballcock into the opening in the bottom of the tank.

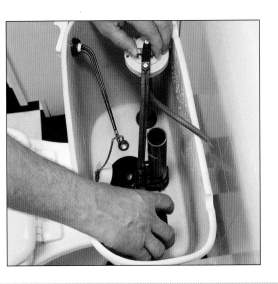

3 **Align the float arm and assemble** Adjust the ballcock assembly so the socket for the float arm is pointing toward the overflow tube. Thread the float ball onto the arm, and the float arm into the ballcock socket. Now twist the ballcock assembly so the float arm is positioned behind the overflow tube. Make sure the float ball does not come in contact with the side of the tank.

4 **Attach the refill tube to the overflow** The new ballcock assembly will come with a new plastic overflow tube. Connect one end to the ballcock by slipping it over the outflow nipple. This is usually just a press-fit. The other end of the tube either will fit into a hole in a plastic cap on the overflow tube or will fit over a tube clip and guide that hook onto the edge of the overflow tube.

5 **Screw the mounting nut to the tailpiece** All that's left to do is thread on and tighten the mounting nut that secures the ballcock assembly to the tank. Tighten with hand pressure at first, recheck the ballcock for proper position, and then finish tightening with an adjustable wrench. Reconnect the water supply line, turn on the water, and check for leaks. Adjust the water level as necessary; *see page 94.*

Servicing
a Flush Valve

1 **Adjust the tank ball** A constantly running toilet can be caused by a poor seal between the tank ball or flapper and the flush valve. First try repositioning the ball so it's directly over the flush valve by adjusting the guide arm that supports the tank-ball lift wire (or by adjusting the lift wires; *see page 95*). Another simple fix is to remove the ball and scrub off any deposits that may have formed on the ball or the seal.

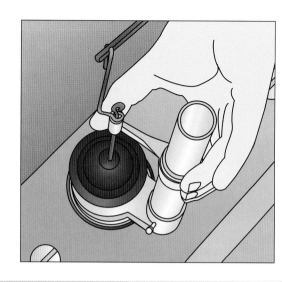

2 **Replace the tank ball** If the tank ball is clean and adjusted properly, it may be cracked or split, or just plain worn out. If your fingers come away black after handling it, it has degraded substantially and needs to be replaced. Most tank balls have a socket on the end that accepts a threaded lift wire. If you're replacing the ball, take the time to remove any deposits on the opening in the flush valve with an abrasive nylon pad.

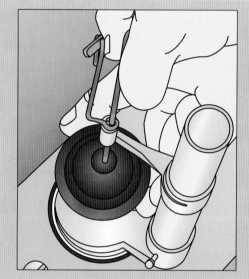

3 **Replace the flapper** If your toilet uses a flapper instead of a tank ball, replacing it is a simple task. First, remove the chain from the lift arm, and then slip each ear of the flapper off its associated lug on the sides of the overflow tube. On flappers that slip on over the overflow tube, remove the refill tube and then slide the flapper up and off the tube. Install a new flapper by reversing the disassembly steps.

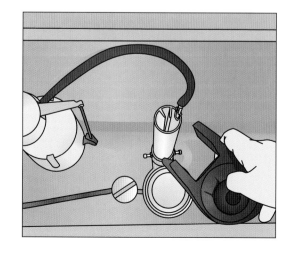

Installing a New Flush Valve

1 **Remove the old valve** To install a new flush valve, start by turning off the water to the toilet; empty the tank and remove it (*see page 105*). With the tank resting upside down on an old towel or blanket, use a spud wrench or a large pair of channel-type pliers to loosen the spud nut. Remove the nut and the old flush valve. Clean the lip of the hole on both sides of the tank.

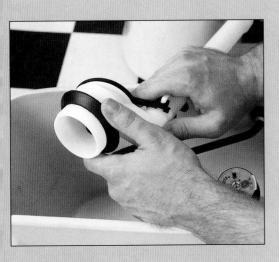

2 **Install the cone washer** Slide the new cone washer supplied with the new flush valve onto the tailpiece so the beveled side is facing the end of the tailpiece. Then insert the flush valve into the tank opening so that the overflow tube is facing the ballcock.

3 **Replace the spud washer** Thread the spud nut onto the tailpiece and tighten it with the spud wrench. Go easy here if the new flush valve is plastic: Overtightening will at best cause parts to warp—at worst, to crack or split. Press a new spud washer over the tailpiece and reassemble the toilet tank, reversing the steps you took to remove it.

Fixing a Leaky Toilet

Potential leak areas There are four main areas for potential leaks on a toilet. From the ground up they are, the seal between the bowl and closet flange, the water supply and shut-off area, the ballcock connection to the tank, and the tank-to-bowl connection. If water is seeping out from under the bowl, try tightening the mounting nuts (*Step 1*) or replacing the wax ring (*page 105*). For the other leaks, try tightening the tank bolts (*Step 2*).

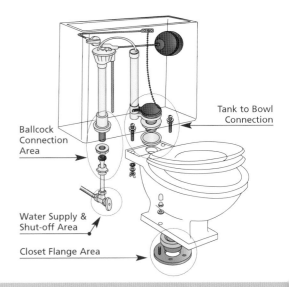

Tank to Bowl Connection

Ballcock Connection Area

Water Supply & Shut-off Area

Closet Flange Area

1 **Tighten the mounting bolts** If water leaks out from under your toilet, the seal provided by the wax ring is no longer intact. If you're lucky, you can stop the leak by tightening the mounting bolts that connect the bowl to the closet flange. A word of caution here: Overtightening these bolts can crack the toilet—if they feel snug, leave them be; instead, replace the wax ring (*see page 105*).

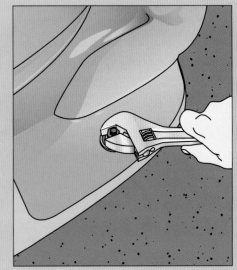

2 **Tighten the tank bolts** If you see water dripping down the back of the bowl, the problem may be a poor seal between the tank and the bowl. Before replacing the spud washer or flush valve (*see page 103*), try tightening the tank bolts. Insert the tip of a screwdriver inside the tank into the threaded head of one of the tank bolts. Hold the bolt still with the screwdriver, and tighten the tank bolt with a socket wrench or an adjustable wrench.

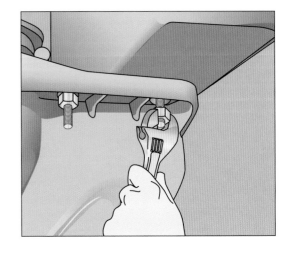

Replacing a Wax Ring

If you've got water seeping out from under your toilet, chances are the wax ring needs to be replaced. Before you do this, try tightening the mounting bolts (*see page 104*). If this doesn't stop the leak, it's time for a new ring. There are four basic steps to replacing a wax ring: removing the toilet and old wax ring, setting the bowl on the new wax ring, connecting the tank, and reconnecting the water supply. When taken a step at a time, this is a simple job as long as you've caught the leak before any permanent damage was done to the floor. If the floor is damaged, you're best off calling in a carpenter to remove and replace the damaged area. (You may also need to replace the closet flange; consult a plumbing professional if you're not sure.) If there is no damage, the biggest challenge you'll face may be removing the old mounting bolts (*see Step 1*). After that, it's clear sailing.

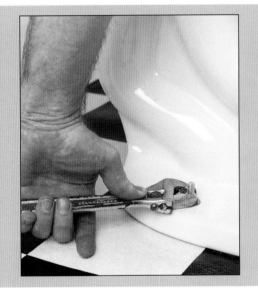

1 Remove tank bolts and trim caps To replace a wax ring, start by shutting off the water to the toilet and emptying the tank completely. Next, pry off the decorative caps at the base of the toilet that cover the mounting nuts. These nuts thread onto the closet bolts and have a well-deserved reputation for not coming off easily. If penetrating oil doesn't help, you'll have to cut the nuts off with a hacksaw.

2 Remove old toilet Once the mounting nuts have been removed, lift the tank gently off the bowl and set it on a towel or blanket to protect it and the floor. Then pull the toilet gently off the floor and set it aside. Since there's sure to be water still remaining in the trap, be careful as you move it about; empty this water into a bucket and set the toilet on its side. If you find the toilet won't come up easily, try rocking it gently from side to side to break the old seal.

3 **Remove the old wax** Plug the drain opening with a damp rag to prevent sewer gas from escaping into the house. Remove the old closet flange bolts from the closet flange, and set them aside if you're planning to reuse them. Then use a putty knife to scrape away the old wax from both the closet flange and the underside of the toilet around the horn (this is a rather messy job). Wipe away any remaining wax with a clean, soft rag.

4 **Install a new wax ring** There are two basic types of wax rings available: a simple ring of wax, and a ring with a rubber no-seep flange. I prefer the no-seep flange ring, since the flange is added insurance against leaks. Insert the bolts in the closet flange, and then position the wax ring on the flange. Apply a coil of plumber's putty around the bottom edge of the base.

5 **Set the bowl in place** Remove the damp rag from the drain opening and position the bowl over the closet bolts and gently lower the bowl; a helper is invaluable in helping line everything up. Once in place, press firmly down on the bowl. Don't stand or jump on it—this will only overcompress the wax ring, resulting in a poor seal. Then thread the mounting nuts on the flange bolts and alternately tighten each nut until the bowl is flush with the floor.

6 **Prepare the tank** With the bowl in place, the next step is to reattach the tank. Flip the tank upside down on the bowl and check the spud washer. If it's worn, replace it; *see page 103.* In some cases, you can get by without replacing the spud washer by applying a coat of pipe-joint compound onto the exposed face of the washer before reinstalling the tank.

7 **Install the tank** Now turn the tank over and set it on the bowl so the spud washer is centered on the inlet opening. Then align the holes in the tank with the holes in the bowl, and insert the tank bolts. This is also a good time to replace the washers on the tank bolts that prevent water from leaking out the holes that the tank bolts pass through.

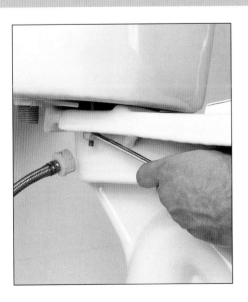

8 **Tighten the tank bolts** To tighten the tank bolts, insert the tip of a long screwdriver in the slot in the bolt. Thread on the nut by hand until it's snug. Then switch over to a socket wrench or an adjustable wrench to finish tightening. Here again, proceed with caution— overtightening can crack the tank. Finally, hook the supply line back up to the ballcock tailpiece, turn on the water, and test for leaks.

Chapter 8
Replacing Fixtures

There are three main reasons to replace a fixture instead of repairing it: It's a quality fixture but you can't find replacement parts anymore, it's a cheap fixture and you want to upgrade it, or you're just after a new look. If the existing fixture is a quality brand (check a plumbing supply house if in doubt) and it's not working properly, I urge you to fix it instead of replacing it. To get the same quality now, you'll have to spend a lot more money than it will cost to repair it.

If the existing fixture is cheap, downright ugly, or not repairable, it's time to replace it. In addition to quality, what you're looking for in a replacement is a fixture that's as close to the original in terms of plumbing connections. The more similar the connections on the new one are to the original's, the easier it is to install.

In this chapter, I'll cover how to replace a variety of fixtures: showerheads (*page 109*); tub spouts (*page 110*); sinks (*pages 111–119*); and faucets, including standard, wide-spread, and tub/shower (*pages 120–125*).

When you go to purchase a replacement, spend the extra money to buy a quality brand name you can trust, like American Standard, Delta, Eljer, Kohler, Moen, or Price Pfister, to name a few. There's a lot of junk out there these days—a lot of plastic chromed housings and stamped metal parts that just won't stand up over time. You'll save money in the long run because you won't be constantly replacing cheap fixtures.

A quality fixture, whether it's a faucet made of brass, or sink of cast iron, will provide years of service. And when it does need repair (and all fixtures will eventually), you'll likely be able to find replacement parts because the company who manufactured it will still be in business.

When it's time to remove the old fixture, go slowly and don't force anything. Over time, plumber's putty can bond a fixture in place with surprising tenacity. Pulling out a sink or faucet without first breaking this bond can cause significant damage to the underlying surface (it's all too easy to pull plastic laminate off a particleboard or plywood base). In most cases, simply running a putty knife under the edge of the fixture will prevent this from happening.

Replacing a Showerhead

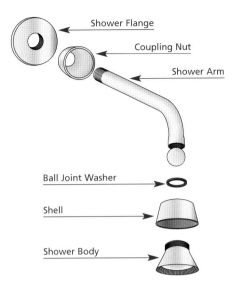

Shower Flange

Coupling Nut

Shower Arm

Ball Joint Washer

Shell

Shower Body

Anatomy Most showerheads are comprised of three main parts: flange, arm, and body. The flange fits snugly around the arm to cover the wall hole. The arm connects to the tube coming up from the shower faucet. In most cases, there's a ball joint on the end of the arm so the body can be adjusted. The body either screws directly onto the arm or is secured by way of a coupling nut.

1 **Remove the old head** There's no need to turn off water when replacing a showerhead; just hold the shower arm firmly with one hand and loosen the body or coupling nut with an adjustable wrench. If it's a direct screw-on type body, there will be flats for a wrench where it connects to the arm. As the head is loosened, be prepared, since any water in the pipe above the faucet will drain out.

2 **Install the new head** Before installing the new showerhead, clean off any old pipe-joint compound or Teflon tape from the threads on the end of the arm. A small brass brush or an old toothbrush works well. Apply fresh Teflon tape to the arm, and thread on the new head, making sure to insert a ball-joint washer if needed. Tighten the body by hand until snug, then give it a quarter turn with an adjustable wrench or channel-type pliers, with the jaws taped to prevent scratches.

Replacing a Tub Spout

1 **Check for setscrew** Before you try to remove a tub
spout, check whether it's secured to the supply nipple
with a setscrew. You'll find the setscrew underneath the
spout near the wall. On some spouts, the setscrew simply
presses up against the nipple, and backing it out a turn or
two will suffice. Another type of spout, common in modern homes, uses the setscrew to tighten or loosen a clamp
that holds the spout in place on the nipple.

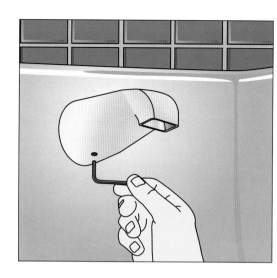

2 **Unscrew the spout** The best way I've found to
remove a threaded-on spout is to insert an old dowel
or a hammer handle of in the end and turn it counterclockwise. Gentle, steady pressure is the way to go. If you
jerk it, you may crack the spout, making it difficult to
remove. If both the spout and nipple end up coming out,
grip the nipple with channel-type pliers or a pipe wrench
and unscrew the spout with the dowel.

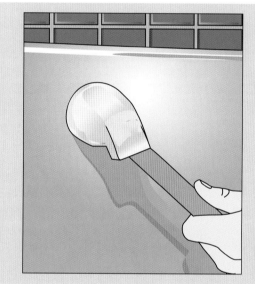

3 **Replace the spout** If the nipple came out with the
spout, apply pipe-joint compound or Teflon tape to
the threads and screw it into the elbow in the wall. Then
wrap a couple turns of Teflon tape around the exposed
end, and thread on the new spout. If you find you have
to unscrew the spout to align it correctly, don't—it'll leak.
Instead, remove the spout, rewrap the threads with
another turn or two of tape, and reinstall it. Newer clamp-
on–style spouts slip over the nipple and are secured by
tightening the built-in clamp with the setscrew.

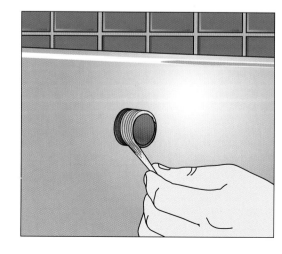

Removing a Sink

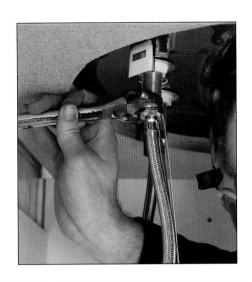

1 **Disconnect the supply** To remove a sink, first turn off the water supply to the faucet. If the sink has shutoff valves, turn the knobs clockwise. If not, shut off the hot and cold water (the main valve and the water heater valve). Open the faucet to drain out any water in the pipes, and then use an adjustable wrench or a basin wrench to loosen the coupling nuts that connect the supply lines to the faucet.

2 **Disconnect the waste line** If the drain hole in the new sink is located where the old drain was, you can get by just loosening the slip nut on the drain tailpiece. Otherwise, you'll need to remove the trap and adjust its position to fit the new sink. With a bucket under the trap and with rags on hand, loosen the slip nuts that connect the trap to the tailpiece and waste line, using channel-type pliers. Then carefully remove the trap and empty it into the bucket.

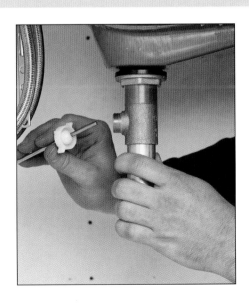

3 **Disconnect the pop-up linkage** If you're planning on installing an exact replacement for a bathroom sink where you'll be reusing the faucet, you'll need to disconnect the pop-up linkage that connects the pop-up handle on the faucet to the drain stopper. Loosen the retaining nut or thumbscrew, and slide the pieces apart. (For more on pop-up linkages, *see page 84.*) If there are any sink-mounting clips, remove them before pushing the sink up and out of the cutout.

Sink Mounting Variations

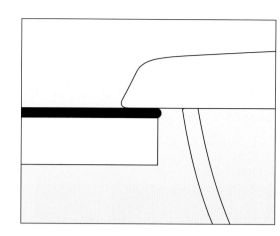

1 **True self-rimming** True self-rimming cast-iron or porcelain sinks are held in place on the countertop with a combination of their significant weight and a thin layer of sealant (such as silicone caulk or plumber's putty). Whichever type you use, the weight of the sink when placed will squeeze out any excess sealant, forming a watertight seal under the small flat on the rim.

2 **Self-rimming with clips** Although it's called self-rimming, this style of sink (the common stainless-steel kitchen variety) really isn't. To create a watertight seal, it requires a dozen or so small retainer clips. These clips hook onto a lip on the underside of the sink and pull the sink down tight against the countertop when the screws are tightened. As with true self-rimming sinks, a bead of sealant or putty is necessary for a good seal.

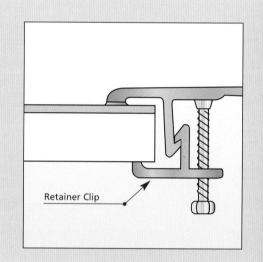

Retainer Clip

3 **Rimless with mounting frame** Before there were self-rimming sinks, the most common style of sink was a rimless sink with a mounting frame. Although it's held in place with clips similar to those used for the sink in Step 2, the difference is that the clips attach to a mounting frame or ring instead of to the sink itself. The advantage to this is that the sink is virtually flush with the countertop. Here again, putty is required for a good seal.

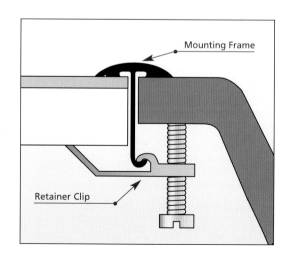

Mounting Frame

Retainer Clip

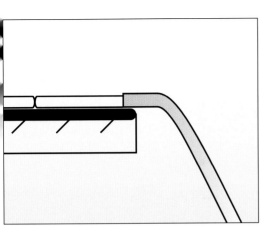

4 **Flush with tile** Installing a sink so it ends up flush with the countertop can also be achieved by first installing the sink and then building up the countertop around it. This is a common way to install a sink where the countertop is to be tiled. The disadvantage here is that the grout surrounding the sink will allow water to seep in, eventually causing the seal between the sink and countertop to fail.

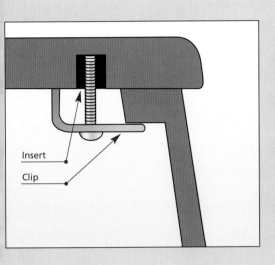

Insert

Clip

5 **Under-counter** Under-counter sinks have become increasingly popular with the advent of solid-surface materials like Corian and Avonite. This style of sink is pressed up under the counter and held in place with clips that screw into inserts embedded in the underside of the countertop. Silicone caulk works best as a sealant here, as it also serves as an adhesive to help hold the sink securely in place.

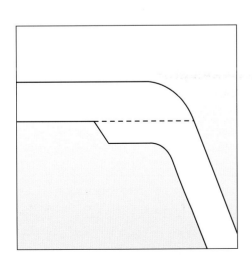

6 **One-piece (integral) sink/countertop** The ultimate solution to stopping water from leaking between a sink and its countertop is to form the sink and countertop as one unit, out of the same material. This eliminates the problem of having to create a seal between two dissimilar pieces. The big disadvantage to a one-piece or integral sink/countertop is that when either the sink or the countertop gets damaged beyond repair, you'll have to replace the entire unit.

Installing a One-Piece Sink/Countertop

Besides the fact that you don't have to worry about water seeping in between the sink and the countertop, the other thing that's great about one-piece sink/countertops is how easy they are to install. There's no fiddling around trying to get the sink to fit into the countertop: Basically you just set the unit on the vanity, secure it, and hook up the water and waste—a weekend afternoon of work. If you're replacing an old sink/countertop with a one-piece unit, try to find one that has the drain hole located in a position similar to your old sink. That way you likely won't have to do much to reconnect the waste, outside of repositioning the trap a bit—here's where flexible waste and supply lines come in handy.

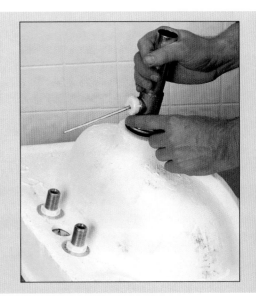

1 **Attach the faucet and tailpiece** Given the opportunity, I'll install a faucet set and tailpiece in a sink before installing the sink—or in this case, the sink and countertop. The faucet and supply-line mounting nuts are just a whole lot more accessible. (If you install the sink first, getting to these will be difficult at best, even with a basin wrench.) Flip the sink upside down and set it on a bench or table protected with an old blanket or towel, and install the faucet and tailpiece.

2 **Apply caulk to the vanity** Although you won't have any leaks between the sink and countertop, it's a good idea to create a watertight seal between the edge of the countertop and the vanity. Without this seal, water spilled over the edge could potentially worm its way into the vanity. To prevent this, run a bead of silicone caulk around the top edge of the vanity before setting the sink/countertop in place. This also forms a bond between the top and vanity to hold it in place.

3 **Install the sink** With the aid of a helper, set the sink/countertop carefully on top of the vanity. Make sure it's centered and the backsplash butts up solidly against the wall. In addition to the silicone, some sink/countertops provide additional means to secure the top, such as threaded inserts embedded in the top. These accept screws driven up through corner blocks in the top edge of the vanity.

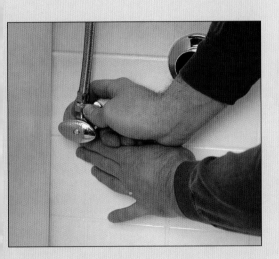

4 **Attach the supply and waste lines** Let the silicone set up for a half-hour or so, then hook up the supply and waste lines. Flexible lines will simplify this task. For the supply lines, wrap Teflon tape around the threads of the shutoff valves, thread on the connecting nuts, and tighten them with an adjustable wrench. For waste lines, position the trap on the tailpiece and tighten the slip nuts. Turn the water back on and check for leaks. Remove the faucet aerator, and run water to flush the faucet.

5 **Caulk at the backsplash** Most one-piece sink/countertops have a built-in backsplash, to reduce the chance of water leaking into the vanity or down the wall. Here again, since it's all one unit, this is very effective. But since you'll inevitably have an occasional splash where water hits the wall and runs down, it's a good idea to seal the any gaps between the backsplash and the wall with a bead of silicone caulk.

Installing a Drop-In Sink

The biggest challenge to installing a drop-in sink in a countertop is accurately marking and then cutting the hole for the sink. A paper or cardboard template will ensure that the hole is both the right size and located in the correct position; a jigsaw or saber saw will make quick work of cutting it out. Self-rimming sinks and rimless sinks with a mounting frame work equally well as drop-ins. In either case, you'll need to create a watertight seal between the sink and the countertop. Plumber's putty is the best choice here. It forms a tight seal but will still allow you to easily remove the sink at a later date. Silicone also forms a good seal, but its higher adhesive properties makes it almost impossible to remove the sink later without harming or destroying the countertop.

1 **Mark the cutout with a template** Many sink manufacturers provide a paper pattern of the hole you'll need to cut out in the countertop. If one is supplied with your sink, cut it out and tape it to the sink so that it's centered from side to side and the specified distance from the backsplash. Then mark around the perimeter of the template with a marker. If there isn't a pattern, make one out of paper or cardboard, using the manufacturer's dimensions.

2 **Cut the hole with a jigsaw** With the sink cutout marked, begin by drilling a ⅜" starter hole for a jigsaw blade inside the cutout, near the line you drew in Step 1. Then use a jigsaw fitted with a wood-cutting blade to cut the hole. Take your time and cut to the inside (or waste portion) of the line. As you near the end of the cut, support the sink cutout with your other hand to prevent splitting or cracking the underlayment and laminate.

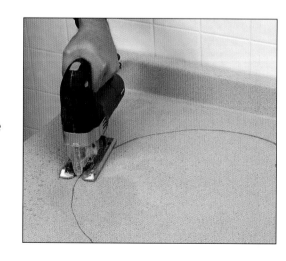

3 **Seal around the rim** If you're installing a self-rimming sink without clips, apply a generous bead of silicone around the edge of the cutout. For most other sinks, apply a continuous coil of ½"-diameter plumber's putty around the edge. Don't scrimp with either material: Any excess will squeeze out and can easily be removed later, and the last thing you want is not enough sealant. As added insurance, apply a coat or two of latex paint to the cut edge to seal it before installing the sink.

4 **Install the sink** Before dropping in the sink, set it upside down on a workbench or table protected with a towel, and install the faucet and tailpiece according to the manufacturer's instructions. I'd suggest getting some help to set the sink in place. Have your helper support the sink from below so that you can concentrate on positioning the sink. If mounting clips are used, install them. Then remove any excess sealant.

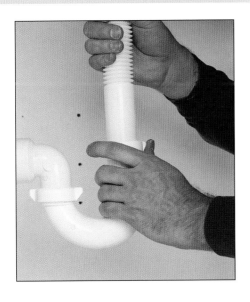

5 **Connect the supply and waste lines** Now you can connect the supply lines to the faucet set and the waste line to the tailpiece. Flexible lines will make both jobs easier. For the supply lines, wrap a couple turns of Teflon tape around the threads of the shutoff valves and tighten the connecting nuts with an adjustable wrench. For the waste line, adjust the trap and tailpiece to fit, tighten the slip nuts by hand, and then a tighten them a quarter turn with channel-type pliers.

Installing a Kitchen Sink

One of the nice things about modern kitchen sinks is that some styles (such as double-basin self-rimming stainless-steel sinks) have been standardized to fit in precut holes in ready-made countertops. If you're installing a new countertop and sink, check to make sure they're made to fit together. If you're installing a sink in a new countertop where you'll have to cut a hole, or enlarge a hole in an existing countertop, follow the procedure on *page 116*. In situations where you will just be replacing a sink, you'll first need to remove the old one; *see page 111* for step-by-step directions on how to do this.

1 **Apply caulk or putty** All self-rimming sinks rely on a sealant to keep water splashed on the countertop from seeping under the rim. Set the sink upside down on the countertop, protected by a towel or blanket. Then apply a continuous, generous bead of silicone caulk or a ½"-diameter coil of plumber's putty around the rim. Alternatively, you can apply the sealant to the edge of the sink cutout.

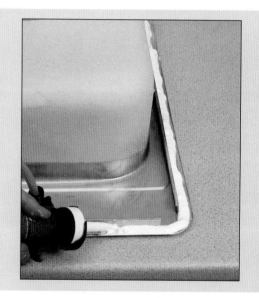

2 **Install the faucet** It's also best to install the sink faucet at this time, as you have better access to the mounting nuts. Slide the sink over on the countertop so it overhangs far enough to allow you to insert the faucet set from underneath. Use the gasket supplied, or pack the cavity with plumber's putty. Secure the faucet by tightening the mounting nuts.

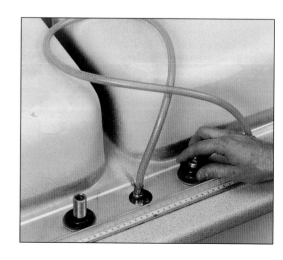

③ Set the sink in position I intentionally didn't install the strainers in the sink in Step 2 because I find it makes moving and positioning the sink a lot easier. Slip one hand into each basin hole (you may want to wear gloves because of the sharp metal edges); then lift the sink and set it in the cutout. It should be a snug fit. Press down to squeeze out any excess caulk or putty; you'll clean up the excess after tightening the mounting clips.

④ Attach mounting clips Most self-rimming stainless-steel kitchen sinks come with a dozen or so mounting clips. Space them out equally on all sides, and then tighten them with a screwdriver (an extra-long screwdriver will be a big help) or a nut-driver. I like to work alternate edges of the sink, tightening a clip on one side, then the clip opposite it. This helps ensure an even squeeze-out of the putty or sealant.

⑤ Remove excess sealant Once the sink is locked in place with the mounting clips, you can remove any excess sealant. For plumber's putty, I run a sharp knife around the edge of the sink and peel off the putty. Silicone caulk is a bit messier. I like to use a plastic putty knife to scrape off most of the excess. Then I follow up with a clean, soft rag to wipe away any remaining caulk. With the sink in place, now you can add the strainers and hook up the waste and supply lines.

Replacing a Faucet

Although it's a simple job, replacing or upgrading an existing faucet does present a challenge. The challenge is accessing the parts—much like changing the oil filter in many cars. Because of the location of the faucet, you'll end up on your back reaching up behind the basin to loosen the mounting nuts. At the same time, you have to navigate past the supply and waste lines, working in an area that provides little if any elbowroom. Fun, huh? It's not really as bad as it sounds. Fortunately, there's a nifty tool that can alleviate much of the problem—it's called a basin wrench, and it's designed for one thing: removing and installing fixtures in sinks (*see Step 2, below*).

1 **Disconnect the supply lines** Before you can remove your old faucet, you'll need to turn off the water and disconnect the supply lines leading to the faucet. If your sink has shutoff valves, turn them clockwise to shut off the water. If not, you'll need to turn one of the main water lines. With the water off, use an adjustable wrench to loosen the nuts connecting the faucet supply lines to the shutoff valves or main water lines.

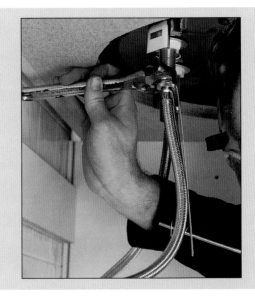

2 **Loosen the mounting nuts** Here's where the basin wrench comes in. Position the jaws of the wrench around a mounting nut, and turn the handle. If the nut doesn't loosen, give the handle a series of quick, hard turns to break it free. If after repeated efforts it still won't budge, you may have to disconnect the waste line and remove the sink for full access to these notoriously stubborn nuts.

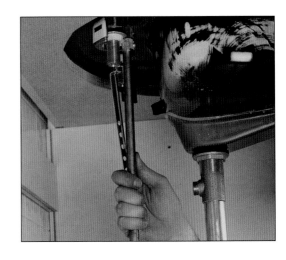

3 **Remove the old faucet** If you're replacing the faucet on a bathroom sink, you'll need to disconnect the pop-up mechanism that controls the drain stopper before you can remove the faucet. (*See page 84* for detailed instructions on how to do this.) Grasp the faucet firmly with both hands and pull up. The putty or caulk used to install the original faucet often develops a surprisingly strong bond over time.

4 **Clean the sink** Once you've pulled out the old faucet, it's important to remove any old putty or caulk from the sink. If you don't, you may not get a good seal under the new faucet. Use a metal or plastic putty knife to scrape away the bulk of the old sealant. Then clean the surface thoroughly with a soft rag and some denatured alcohol.

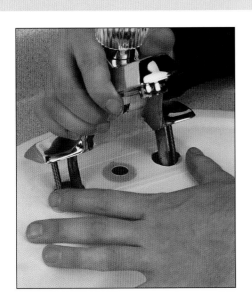

5 **Install the new faucet** Use the gasket provided with the new faucet, or pack the cavity with plumber's putty. If possible, attach flexible supply lines to the faucet first, then set it in position and thread on the mounting nuts from underneath by hand; use the basin wrench to finish tightening them. Wrap fresh Teflon tape around the threads of the shutoff valves of water supply lines, and tighten the connecting nuts with an adjustable wrench.

Replacing a Wide-Spread Faucet

Unlike its one-piece cousin, the wide-spread faucet is made up of a number of parts, typically a spout, separate handles, and options such as a sprayer. These faucets can be installed in any sink designed for a one-piece faucet. Since there are multiple parts, they are a bit more complicated to install because the individual pieces have to be connected with flexible tubing. On some faucets, a T connector hooked up to the spout accepts hot and cold water from valves installed beneath the faucet handles. Other systems run hot and cold water directly into the handle, which then routes the mixture off to the spout via a flexible line. Whichever system you have, you'll need to first remove the existing faucet before installing the new one; see *page 120* for directions on how to do this.

1 **Install the spout** After you've removed the old faucet and cleaned off any putty or caulk residue, you can start installing the separate parts of the wide-spread faucet. Install the spout first if it uses a T connector so that you can position it properly. Use the gaskets supplied with the faucet for a watertight seal. Or if you prefer, discard the gaskets and pack the cavities with plumber's putty.

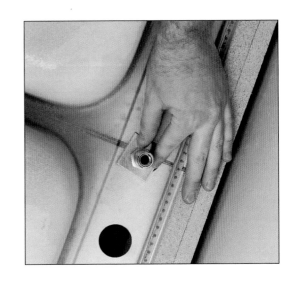

2 **Attach the flange with a nut** The spout on a wide-spread faucet is held in place by way of a flange and a nut. With the spout held firmly in place, thread on the washer and nut supplied with the faucet. Then tighten the nut with an adjustable wrench. Some manufacturers supply a special "socket" that can be used with an ordinary screwdriver to tighten the nut.

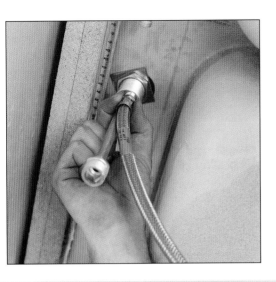

3 **Attach the valves** If the wide-spread faucet you're installing uses faucet valves to route the water to the spout, they can be attached next. Some types are inserted up throughout the sink; others go in from above. Use the mounting nuts supplied by the manufacturer to secure the faucet valves. Tighten them firmly with a large adjustable wrench or channel-type pliers. Then wrap each valve outlet with Teflon tape, and thread on the flexible supply lines provided.

4 **Install the T connector** If a T connector is used to connect the spout to the faucet valves, wrap a couple turns of Teflon tape around the threads of the spout and thread on the T connector. Use an adjustable wrench to tighten it in place. Then wrap Teflon tape around the open ends of the T connector, and thread on the flexible lines coming from the faucet valves that you hooked up in Step 3.

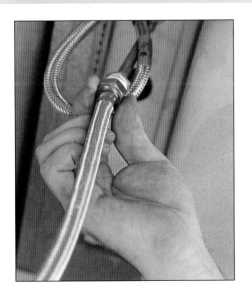

5 **Connect the supply lines** Finally, connect water to the faucet valves. The easiest way to do this is to run a pair of flexible supply lines between the shutoff valves and the faucet valves. Wrap Teflon tape around the threads before attaching the connecting nuts. Attach the faucet handles (and flanges, if used) to the faucet valves with the screws provided. Snap in trim caps if supplied.

Replacing a Tub/Shower Faucet

Replacing the faucet for a tub/shower unit is a challenging project. If you're comfortable with sweating copper pipe and working in tight spaces, it may not be more than an afternoon of work. If you're not and you jump into this, you may end up having to call in a professional. The lack of space to work in is really the biggest challenge. If you're lucky and there's an access panel behind the faucet body, or if the area behind the shower wall is unfinished (as in a basement), the job will be a lot simpler. If there isn't an access panel, you'll be confined to the opening hidden by the escutcheon. Before taking on this job, I'd suggest removing the escutcheon (*see Step 1*) to see what kind of space you have. The other complication is that you have to deal with four pipes that all run into one faucet body: the inlets for hot and cold water, and the outlets to the tub spout and showerhead.

1 **Remove the escutcheon** Turn off the water supply, and drain any water from the lines. Next, remove the control handle by first prying off the small plastic or metal trim cap covering the handle screw; then remove the handle screw and pull off the handle. Now remove the screws holding the escutcheon in place. Pull the escutcheon gently away from the shower wall to prevent tearing the gasket that forms the seal between the wall and the escutcheon.

2 **Remove the old faucet set** To remove the old faucet, you'll need to cut the inlet and outlet lines connected to the faucet. If you've got access from behind, this can be done with a hacksaw. If you don't have access, use a mini-hacksaw. Cut roughly 1" away from the faucet body. This will leave you sufficient space to sweat the unions to reconnect the lines.

3 **Unscrew the faucet body** The faucet body should be attached with screws to a wall cleat running between the wall studs. Remove the screws and lift out the faucet body. Buy a direct replacement if possible, or a new one that's similar in size, to make installation of the new faucet as simple as possible.

4 **Reconnect the pipes** Cut four short lengths of copper pipe to connect the new faucet body to the old lines. A bit of trial-and-error will be needed here to get all the pieces to fit together nicely. After you've tested the fit, ream out the ends of the pipe, clean all the parts, and apply flux as described on *pages 37–38.* Carefully position the faucet body in the opening, using unions to connect it to the existing pipes.

5 **Sweat the pipes** Remove any rubber parts from the faucet body to prevent them from melting when you sweat the joints. (For detailed instructions on how to do this, *see pages 38–39.*) Make sure to protect surrounding surfaces with a heat-resistant cloth or a piece of sheet metal. When everything has cooled, turn on the water and test for leaks. If everything is okay, install the escutcheon, the handle, and the trim cap.

Glossary

ABS (acrylonitrile butadene styrene) – a rigid, black plastic pipe used for drain, waste, and vent systems; it requires its own cement, and it should never be mixed with PVC.

Aerator – a device that snaps or screws onto a faucet spout and creates bubbles in the flowing water to prevent splashing.

Auger – a tool used to clear drains or toilets; turning a crank on the auger causes a coiled wire to spin and auger or "drill" through solid waste.

Ballcock – a valve inside the toilet tank that automatically fills the tank to a preset level when the toilet is flushed; there are three common types of ballcocks: diaphragm, plunge-valve, and float-cup.

Ball-type faucet – a type of faucet that uses a rotating ball to control water flow and temperature.

Cartridge faucet – a type of a faucet that uses a replaceable cartridge to control the flow and temperature of water; easily identified, since the cartridge moves up and down with the handle.

Cast-iron pipe – large, heavy pipe that's used for waste and vent systems in older homes; it has been replaced largely with ABS and PVC.

Clean-out – a fitting with a removable plug that provides access to the drain system to allow you to remove obstructions with a snake or auger.

Closet flange – a fitting that attaches to the floor and accepts the toilet; slots in the flange hold flange or "Johnny" bolts, which pass up through the base of the toilet to hold it in place.

Compression faucet – a type of faucet that uses a stem with a washer on the end to control the flow of water.

Compression fitting – a fitting used in situations where the connection may need to be taken apart later; a nut forms a seal by forcing a compression ring against a pipe.

Copper pipe – rigid, thick-walled pipe made of copper, used primarily for hot and cold water distribution inside a home.

Coupling – a fitting that's used to connect two lengths of pipe in a straight run.

CPVC (chlorinated polyvinyl chloride) – a type of plastic pipe that is approved for both hot and cold water distribution—unlike PVC, which is approved to carry only cold water and waste.

Dielectric union – prevents corrosion when joining dissimilar metal pipe, like copper pipe to steel pipe; one end is threaded for the steel pipe, the other end is soldered onto the copper pipe.

Disk faucet – a type of faucet that uses a disk assembly to control the flow and temperature of water; as the handle is turned, the disk assembly rises and falls and breaks contact with a spring-loaded seat.

Diverter valve – usually contained within the faucet body, this diverts water to the sprayer when the sprayer handle is activated.

DWV (drain-waste-vent) – the system that carries away liquid and solid waste; allows sewer gas to escape safely and prevents water from siphoning out of traps.

Elbow, or L – a fitting that connects pipe to pipe at angles to each other; may be 45° or 90°; can have two female ends, or one female and one male (known as a "street" fitting).

Escutcheon – a decorative trim piece that fits over a faucet or pipe.

Fitting – any device that connects pipe to pipe or pipe to a fixture.

Fixture – any permanently installed device connected to the water supply system (such as a bathtub, toilet, sink, or faucet).

Flare fitting – a fitting most often used with flexible copper pipe for LP gas lines when the connection may need to be taken apart; a special flaring tool shapes the pipe end to fit over a matching flare inside the fitting.

Galvanized-iron pipe – very strong, heavy pipe threaded on both ends; commonly used in older homes for water distribution; corrodes over time and has been largely replaced with copper pipe.

Gate valve – uses a regulating mechanism similar to the gate in old irrigation systems—the gate is raised to allow water to pass; typically used as the main water shutoff in older homes.

Globe valve – uses a stem-and-washer setup similar to a compression faucet to regulate the flow of water; more reliable than a gate valve, and easily repaired.

Hose bib – a type of valve with an externally threaded outlet to accept a hose fitting; found outdoors for hoses and in laundry rooms for connection to a washer.

Main shutoff valve – a valve located between the water service coming into the house and the interior water distribution system that's used to turn off water to the house.

Main vent stack – the principal artery in the drain-waste-vent system that all branch lines connect to.

Nipple – a short piece of pipe with male threads that's used to join two fittings.

Packing – a fibrous or rubber material that prevents pressure leaks; typically found on old valve-and-stems of compression-type faucets.

Pipe-joint compound – a sealant with the texture of toothpaste that is applied to female threads to create a watertight joint.

Plastic pipe – a generic term used to describe any of the following types of pipe made of plastic: ABS, CPVC, PB (polybutylene), PE (polyethylene), and PVC.

Plumbing code – regulations adopted by the local community to control any plumbing work done in the community.

Plumber's putty – a dough-like sealant used to prevent gravity leaks; typically used on drains, under faucets, and under sinks to prevent water from seeping in.

Pop-up mechanism – a mechanism found on most bathroom faucets that raises and lowers the drain stopper when the handle is activated.

P-trap – a curved drainpipe that attaches to a fixture (such as a sink or bathtub); traps water to create a seal to prevent sewer gas from passing into the home.

PVC (polyvinyl chloride) – a rigid plastic pipe designed for cold-water use; it has virtually replaced cast-iron pipe for DWV systems in new construction.

Reducer – a fitting that allows connection of pipes of different diameter.

Sanitary fitting – a drain fitting with a smooth interior and gradual curves, which allows waste to pass through without clogging.

Snake – a slang term for a drain auger; either manually operated or powered by electricity.

Soldering or sweating – a method of using heat to join together copper pipe and fittings.

Stem – a part of a faucet that rises and falls when the handle is turned; a washer on the end of the stem presses against a seal to control the flow of water.

Supply line – any piping that is used to distribute water (either hot or cold) within a home.

Teflon tape – a thin membrane that's wrapped around male pipe threads to create a watertight joint.

Transition fitting – any adapter used to connect two dissimilar materials (such as steel to copper, copper to plastic, etc.)

Trap – a curved device that allows water and waste to pass through while blocking the flow of air and gas from the opposite direction.

Valve – a device that controls the flow of water through a pipe.

Vent – a pipe that allows air to flow into, and gas out of, the DWV system; prevents water from siphoning out of traps.

Wax ring – a donut-shaped ring of wax that fits between the base of a toilet and the closet flange to create a watertight seal.

Index